# DECODING
## THE LANGUAGE OF GOD

C en

# DECODING
## THE LANGUAGE OF GOD

### CAN A SCIENTIST REALLY BE A BELIEVER? A GENETICIST RESPONDS TO FRANCIS COLLINS

*George C. Cunningham, MD, MPH*

59 John Glenn Drive
Amherst, New York 14228-2119

Published 2010 by Prometheus Books



Inquiries should be addressed to
Prometheus Books
59 John Glenn Drive
Amherst, New York 14228–2119
VOICE: 716–691–0133
FAX: 716–691–0137
WWW.PROMETHEUSBOOKS.COM

14 13 12 11 10 5 4 3 2 1

Library of Congress Cataloging-in-Publication Data

Cunningham, George C., 1930–.
Decoding the language of God : can a scientist really be a believer? : a geneticist responds to Francis Collins / George C. Cunningham.
p. cm.
Includes bibliographical references and index.
ISBN 978–1–59102–766–9 (pbk. : alk. paper)
1. Collins, Francis S.—Language of God. 2. Religion and science. I. Title.

BL240.3 .C86 2009
215—dc22

2009032863

Printed in the United States of America

*To all those who have the desire for truth,*
*the commitment to find it,*
*the wisdom to recognize it,*
*and the courage to accept it.*

# CONTENTS

# ACKNOWLEDGMENTS

I would like to thank Gina Gold for her editorial and typing assistance and her creative suggestion for a title, "The Language of God, Lost in Translation," which I still like. Thanks to my son Mike who helped compensate for my limited computer skills. I also want to express my appreciation to my agent, Claire Gerus, and my editor, Steven L. Mitchell, who provided professional guidance without interfering in any way with the personal expression of my basic message. Finally, thanks to Susan Berman, who did much of the typing and formatting of the notes.

# INTRODUCTION

*The first demand any work of art makes of us is surrender. Look. Listen. Receive. Get yourself out of the way. There is no good asking first whether or not the work before you deserves such a surrender, for until you have surrendered you cannot possibly find out.*

—C. S. Lewis, *An Experiment in Criticism*

## WHY I WROTE THIS BOOK

For most of my adult life I have been immensely interested in answering the ultimate questions. What is the point of it all? What should I do with my life? Where did this universe come from and what is my place in it? In pursuit of answers, I have explored religious, philosophical, and scientific approaches. Through these explorations I have concluded that science and religion as worldviews are basically and fundamentally incompatible.

I was therefore interested in a popular and much-cited book that claimed scientists can believe in a personal God and still be rational scientists. The book, by Francis Collins, is *The Language of God.*[1] Francis Collins was recently appointed director of the National Institutes of Health and was head of the team that mapped the human genome. This remarkable accomplishment made Dr. Collins a media superstar and an internationally known scientist. His book was a best-seller and rated number one in both the science and the religion categories on Amazon.com. He has been interviewed many times on CBC Radio and NPR and on television, including CNN, PBS, *Charlie Rose*, and *The Colbert Report*, and in magazines like *Time*, *Discover*, and *BusinessWeek*. Google lists over 748,000 citations and

Yahoo! 19,300,000 for Dr. Collins. For this reason, his book, in which he claims that the scientific worldview can be accepted without having to give up the truth of his Christian religion, has received extensive promotion and coverage. Prominent Christians have praised the book as a lucid, compelling harmonization of science and faith. The Internet is full of listings and articles referencing the religious sentiments of this famous scientist.

However, after a careful reading of *The Language of God*, I found Collins's arguments and "evidence" that religious beliefs can be reconciled with scientific truth unconvincing. My response in this book is motivated by my strong belief that these arguments need to be critically analyzed and refuted. It is my intention to state Collins's arguments and evidence fairly and completely, not to attack a straw man of my own creation. Wherever possible, I use his own words rather than paraphrasing his concepts. My book follows and responds to each of his chapters with which I have a problem. I focus on the evidence that Collins uses to support his belief that Jesus Christ is the creator god who desires fellowship with humankind.

Members of the religious community, in order to justify their particular beliefs as consistent with science, rationality, and logic, overuse and abuse Dr. Collins's scientific reputation. In citing the geneticist Collins, they ignore the fact that other exceptional geneticists such as James Watson and Francis Crick—the Nobel Prize–winning discoverers of the structure of deoxyribonucleic acid (DNA)—as well as most of the genetic fraternity are atheists or agnostics.

Collins's book attempts to convince readers of two propositions: first, it is rational to believe in a personal God who desires fellowship with humans, and second, this personal God is the historical Jewish teacher, Jesus. I intend to show how Collins's attempt fails and to demonstrate that no one can simultaneously accept the belief in a personal God and still claim to be a logical and rational scientist without engaging in magical thinking.

There are other books by scientists who have claimed to reconcile science, mainly evolution, with their theistic beliefs.[2] In answering

Collins's arguments I will have addressed almost all of the points raised by these authors. I selected Collins as the spokesperson for "believing" scientists because he uses the same basic format as most other believing scientists in trying to justify an irrational belief in a personal God. These Christian scientists state some fact, observation, or problem about the universe and ask for an explanation or solution. They then claim that there is no satisfactory natural explanation or solution and, therefore, they propose a supernatural one.

Examples of this attempt to fill in gaps of scientific knowledge with a "God of the gaps" include the following arguments: (1) humans are moral, and science cannot explain morality; therefore, God is responsible for making human morality, or (2) events called "miracles" occur, and science cannot explain miracles; therefore, miracles are proof of God's action in our universe. By analyzing Collins's arguments in detail, I believe it will be clear that the arguments he uses are not rational, logical, or consistent with modern science. In truth, they are rationalizations for blind, unsupported faith.

I will primarily respond to Collins, but because he so frequently refers to the writings of Clive Staples Lewis (1898–1963), I will also respond to Lewis. C. S. Lewis was an English writer who "found" Jesus and became an ardent defender of the faith in metaphorical children's stories such as *The Chronicles of Narnia* and academic books such as *Mere Christianity*. We need to be clear about what kind of God Collins and Lewis want us to believe in. There is an important distinction between an impersonal, spiritual, creative force with no special relationship with humans and a personal God like Jesus or Allah or Yahweh, who has rules for human behavior and interacts with humankind. Collins makes it clear that he finds no conflict between his work as a "rigorous scientist" and his belief in a personal God, "a God who is unlimited by time and space, and who takes personal interest in human beings."[3]

## A NOTE ON LANGUAGE

In writing this book, I have decided to depart from the usual convention of referring to the deity using the masculine gender. It is clear that a supernatural being does not engage in sexual reproduction. While a recent poll shows that less than 10 percent of Americans believe that God has a face, arms, legs, or other body parts, about four in ten believe that the deity is male.[4] I cannot imagine how this group of believers is using the term *male*, since it cannot mean that God has a penis. I hope it does not imply that god has "uniquely masculine" qualities such as being strong or intelligent, which implies that females do not have the same quality or quantity of these "male" features. One of my objections to all the major religions, and Christianity in particular, is the open and hidden sexism that persists among their defenders and practitioners. What justifies our making God in our male image? If God is a supernatural being, why arbitrarily assign gender, which is characteristic of natural beings?

Whatever you think about the nature of God, the deity must be beyond human limitations of space, time, personality, and sex. While Jesus was a male (the Bible refers to his circumcision), we have yet to establish in our review of the evidence Collins cites that Christ was more than a human being. We are, at least temporarily, suspending any belief that Jesus was God. I am not discussing the reasons that God would choose to take on a male rather than a female form. I leave it to you to speculate on the implications of that divine decision.

I considered referring to the transcendental spiritual being we call God as "it," to avoid the confusing and intellectually dishonest habit of using the masculine to debate the existence of a sexless supernatural being. The pronoun "it" refers to a sexless, nonliving object and might be appropriate since God is not a living being. But I do not intend to be needlessly provocative, so I will simply refer to it as God. I will follow the convention of using *god*, lowercase, to refer to a general supernatural entity, and *God*, uppercase, to refer to the personal God of Collins, namely, Jesus. Of course, I will not change the masculine reference when quoting others.

## ARE SCIENTISTS BELIEVERS?

No, they are not. In his introduction, Collins attempts to show that many scientists agree with him by claiming that scientists' belief in a personal God is more prevalent than many realize.[5] He continues to cite a study in 1997 that found 40 percent of scientists expressed such a belief. And while he doesn't give a reference, I believe it is an article by Edward Larson and Larry Witham in *Scientific American*.[6] How this study is interpreted depends on your point of view. It would seem to show that a majority of scientists, 60 percent, did not believe in a personal God. Among the 40 percent who did, it is not clear that they all believed in the same personal God, namely, Jesus, Allah, or Yahweh. These were American scientists, and as scientists, were much less likely to believe in a personal God than the average American. (For comparison, a recent poll found some 57 percent of nonscientist Americans believe in a personal God, that is, one that controls what happens on Earth.)[7]

These results quoted by Collins were based on responses from ordinary, randomly chosen scientists. Larson and Witham also questioned a group of leading scientists representing the membership of the National Academy of Science.[8] They asked, "Do you believe in a God in intellectual and affective communication with humankind?" This study of top scientists' religious beliefs gave the following results when compared to previous surveys:

| **Belief** | **1914 (%)** | **1933 (%)** | **1998 (%)** |
|---|---|---|---|
| Believe in personal God (believers) | 27.7 | 15 | 7.0 |
| Do not believe (atheist) | 52.7 | 68 | 72.2 |
| Doubt or don't know (agnostic) | 20.9 | 17 | 20.8 |

This study shows 93 percent of top scientists are atheists or agnostics and indicates an increase in atheism and agnosticism as time goes by.

These data, although still cited by even the atheist literature, are now out of date. The most recent study of this question, funded by the

Templeton Foundation, was by Elaine Howard Ecklund and Christopher Scheitle.[9] They collected answers to thirty-six questions from 1,646 randomly selected scientists at twenty-one elite universities. Their results, published in 2007, refute Collins's assertion that up to 40 percent of scientists today believe in a personal God. Of the 1,646 scientists surveyed, 9.7 percent were certain of the existence of God, while 31.2 percent were atheists, and 31.0 percent agnostic. The remaining 28.1 percent either had doubts or believed in an impersonal "higher power." Only 3.7 percent responded that "there is the most truth in one religion," and 22.8 percent agreed that there is very little truth in any religion. When compared to the general public in America, where only 14.2 percent claim no religious affiliation, they found 51.8 percent of scientists have no religious affiliation. (A 2007 survey reports 16.1 percent of the general public is now unaffiliated.)[10]

In terms of translation of religious beliefs into practice, 76.8 percent of the scientists did not attend a religious service or did so less than five times in the last year as compared to 46 percent of the US population. Only 8.6 percent of scientists, compared to 26 percent of the general population, reported attending at least once a week.

Collins is mistaken in his assessment of how many of his fellow scientists would agree with him that Jesus is their own personal God. Collins is definitely in a small minority as a believing scientist. This fact, while far from proof, is consistent with and supportive of my contention that science is logically incompatible with faith in a personal God.

## WHY IS SCIENTIFIC ACCEPTANCE SO IMPORTANT TO BELIEVERS?

There is great respect for the intelligence and accomplishments of individual scientists and for the success of science as an institution in providing reasonable and useful interpretations of our universe. There is no Jewish or Christian or Islamic way to collect data from a radio telescope or build a high-definition TV or predict the weather. The universal application of scientific theories to explain nature has been one of the most productive uses of the human mind. Scientific truths are demonstrated in the technology that has taken man into the vast reaches of space, produced dramatic improvements in diagnosis and treatment of diseases, and supported the electronics revolution. Scientists know a lot about our world.

This respect for the power of scientific knowledge explains why people who wish to defend their belief in God always cite the very small number of scientists who are believers. They use quotations from unbelieving scientists inappropriately to buttress their defense of theism. The believers reason that if scientists can share their beliefs, then those beliefs must be rational and scientifically sound. For this reason, whenever any recognizable scientist expresses some kind of theistic belief, it becomes newsworthy. Any study that states scientists are believers will receive wide distribution in popular and religious media.

Believers are grasping at straws in a vain attempt to win the same respect and esteem for religion that we rightly show science. Defenders of religion constantly assign belief to prominent scientists such as Albert Einstein or Stephen Hawking by quoting them every time they happen to use religious language or metaphor, such as the "mind" or "hand" or "language" of God.

Almost always, these quotes are out of context or misquotes and they ignore these same scientists' clear statements of disbelief. This kind of misrepresentation tends to discredit believers as being interested in winning a public-relations war rather than marshalling logical, rational evidence. Citing quotations from, or making lists of, believing or atheistic scientists is an idle exercise proving nothing. The merits of the arguments concerning the existence of a personal God do not depend on the authority of the persons making them, whether they are scientists or nonscientists. The truth of science and religion is a matter each of us can determine based on reasonable analysis of the evidence and our personal experience.

## SCIENCE VERSUS RELIGION

The persistence and growth of atheism and agnosticism among scientists is especially remarkable given the tremendous efforts and resources religious groups devote to seducing scientists to join them and to neutralize or discredit science that they regard as dangerous to their religious views. Scientists are well aware that they work in institutions that depend on public support, public and private grants, and charitable giving. They are also aware that atheists are the most despised and stigmatized group in the United States.[11] Would you vote for a godless (read: wicked, untrustworthy) atheist for president? Would you want your daughter or son to marry an atheist? Discrimination against atheists is not only tolerated; in some cases it is also legally condoned.

The status of nonbelievers is not helped by their most aggressive spokespersons (Richard Dawkins, *The God Delusion*; Sam Harris, *Letter to a Christian Nation*; Daniel Dennett, *Breaking the Spell*; Christopher Hitchins, *God Is Not Great*),[12] who characterize religions as ignorant and/or evil. The defenders of the faith constantly use adjectives such as "angry," "despicable," "irrational," "dangerous," "God-hating," and, Collins's favorite, "obnoxious," to describe atheists. Given this stigma, scientists have little to gain and much to lose by acknowledging their lack of belief or by actively engaging in the science-versus-religion debate.

The John Templeton Foundation has assets of $1.1 billion and spends $60 million a year to fund scientists and organize workshops that bring together religious spokespersons and scientists to reconcile inconsistencies and support research into "spiritual realities."[13] The Templeton Prize is designed to be larger than the Nobel Prize in terms of dollars. This has the effect of silencing some critics or possibly inducing some into changing their beliefs. In addition, the Discovery Institute, a group dedicated to theistic opposition to evolution, has funded $9.3 million in public-relations programs to discredit scientists and atheism. "As of 2004, the US market for creationism was at least

$22 million as measured by adding up donations to and purchases of products and services from ten of the largest creationist groups."[14]

The Alliance Defense Fund is an organization of over nine hundred believing lawyers, each of whom contribute 450 hours of unpaid legal services annually valued at least $5 million. The core funding of $20 million comes from corporate and individual donations and is devoted to using the legal system to advance religious doctrines. The Thomas Moore Law Center is a similar organization with about three hundred volunteer attorneys that is also supported by corporate foundations and individual gifts. They constitute a potential legal threat to science and individual scientists. They have taken legal action to censure the teaching of evolution and against stem cell research and sent intimidating letters to scientists. This list of resources does not include the hundreds of millions of dollars spent by religious denominations in promotion of their theistic message. Thanks to Gideon International, you will find a Bible in almost every hotel or motel in the United States. In the face of this challenge and the unequal resources of the two sides of the debate, it is amazing that the studies still show that the vast majority of scientists, especially at top levels, remain unbelievers.

## ARE YOU A BELIEVER?

Many human beings have tried to answer these kind of ultimate questions for themselves. But for many of you readers, these questions have been answered for you. When you were born your parents, siblings, and relatives taught you the answers they had learned from their parents. You were born into a family that accepted and lived by a set of beliefs usually adopted without much criticism or questioning. If your parents were religious, long before you reached the age of reason they taught you the ritual, beliefs, and practices of their religion. They enjoyed your unquestioning trust, because they loved you and cared for you. Or perhaps you adopted your religious worldview from a teacher or other religious expert, whom you felt knew more about the details and evidence than you. Perhaps you found religion when in a time of personal distress you found comfort from a religious person or her message. For many people, the truth of this way of thinking, once acquired, is never subjected to examination or critical questioning but only to modification.

In spite of a great deal of religion switching, some of which is related to marriage, most people remain affiliated with the religion in which they were raised. They have a very strong emotional attachment to these long-held beliefs because they provide clear guidance and purpose for their lives. The complete faith that their beliefs are always true provides them with comfort and security. People who share the same beliefs form a supportive community and give one another a sense of belonging to something bigger than themselves. I know this because I was a believer at one time.

People believe what is consistent with their existing beliefs, that is, what they want to believe. This is why a majority of Muslims worldwide do not believe that Muslims were the perpetrators of the horrors of 9/11, and why some Americans believe that the CIA or Jews or Saddam Hussein were responsible in spite of strong evidence to the contrary. Repetition of the truth and attacking the creditability of false explanations are only partially effective once belief is established. This accounts for the persistence of errors in the popular imagination.

Yet, I ask these readers to do something that may be impossibly difficult: specifically, to set aside a lifetime of cherished beliefs for a few moments and approach the discussion in this text with as open a mind as they can. To seek true knowledge, one must question the unquestionable and challenge the unchallengeable. The result may be that you could change or abandon your current beliefs or you could find them strengthened and expanded.

In contrast to some best-selling critiques of religion—such as Richard Dawkins's *The God Delusion*, Christopher Hitchens's *God Is Not Great*, and Sam Harris's *Letter to a Christian Nation*—which unabashedly attack religions and their followers, I feel no need to denigrate all believers. The "militant atheists" do not give religion enough credit for its social contributions. While I deplore the overall effect of theist beliefs on the health and happiness of millions of human beings, I do not think most believers have evil intentions. For the most part they are unaware of just how their belief in the primary importance of a possible supernatural world contributes to the harm done to real people in this real material world.

Atheists and secular humanists praise believers when they tend to the sick, comfort the grieving, and give food and shelter to the unfortunate. But they criticize believers when they threaten the physical and moral health of women by restricting access to voluntary abortion, contraception, vaccination against cervical cancer, and accurate sex education in their quest to stamp out sin. Whether discussing the destruction of the World Trade Center, or the Inquisition, or the hanging of Quakers by Baptists, the words of the seventeenth-century philosopher-mathematician Blaise Pascal come to mind: "Men never do evil so completely and cheerfully as when they do it from a religious conviction."

## REFERENCES TO GOD

God goes back a long way in human history. Anthropological evidence suggests that before the arrival of *Homo sapiens*, the genus *Homo* felt the presence of spirits, unseen forces that inhabited animals and inanimate objects. The first religions were polytheistic, with many different gods that frequently had characteristics of superheroes or were combinations of human and animals. For a very readable history of how monotheism battled against and finally, at least in the West, conquered polytheism, read Jonathan Kirsch's book *God against the Gods*.[15]

Human concepts and the words that describe them change with time. So it is with the word *god*. This makes any logical debate about the reality of the concept difficult even if one makes every effort to be precise and accurate. The task of defining what is meant by the word *god* needs to be completed and the definition accepted in order to defend the existence of the concept. As an unbeliever I do not want to be accused of misrepresenting the nature of the God the believer wants to defend. It is intellectually dishonest to create a straw man so that you can easily knock it down.

The most recent attempt to determine in more detail what the word *god* means to Americans was funded by the proreligion Templeton Foundation.[16] The 2006 survey, which was conducted by Gallup and analyzed by the Baylor University Institute for Studies of Religion, found, as might be expected, that although 91.8 percent believed in a higher power or cosmic force, when it came to describing the nature of god, there were many differences in the responses. The survey underestimated what I would define as unbelievers. They found 10.8 percent were unaffiliated. When broken down by age, 18.6 percent of younger (18–30), 11 percent of middle-aged (31–64), and 5.2 percent of elderly (65+) people were unaffiliated, which may account for the subsequent increase report in more recent polls. The survey also has a plus-or-minus error of 4 percent.

The reliability and objectivity of this report has been sharply challenged by Gregory Paul in an analysis performed for the Council for Secular Humanism.[17] The responses of believers were somewhat arbitrarily sorted into four groups. Since the four groups have some overlapping views, the total does not add to 100 percent. The first group, representing 24.4 percent, sees god as an impersonal cosmic or creative force behind the laws of nature. The largest group, 31.4 percent, are fundamentalists who see God as an angry and judgmental person who interacts with humans. They accept the Bible as accurate and believe in prayers and miracles. Another group describes God as having a critical personality toward sinful humans but only interacting with them on the final judgment day. The final grouping, 23 percent, see God as the author of natural and moral law but also as a comforting and guiding spirit that responds to prayer, occasionally intervenes in miraculous ways, and as someone who, while more forgiving than the God of the previous two groups, still dispenses rewards and punishments after death.

Collins's definition of God includes the following: God is the creator of the universe who desires a special fellowship with humans and interacts with them in accordance with a divine plan including miraculous interactions. God is the eternal embodiment of goodness and took human form as Jesus Christ, who was crucified, rose from the dead, and will award supernatural rewards or punishments to every human being at the time of death.

Throughout this book I use data from polls and surveys, so I want to comment on their accuracy. Responders tend to want to give the "right" answer when asked questions about God. This tends to underestimate the number who are unbelievers, atheists, agnostics, and functional atheists who behave as if they were atheists in spite of some fictitious association or affiliation with a religious group. The number of unbelievers is estimated today at between 16 percent (Pew) and 21 percent (Harris), which means over 60 million unbelievers in the United States.

Some believers may claim that my arguments are not applicable to their authentic Christianity but are aimed at a perversion of Christianity. I reject this allegation. It is true that Christianity is a moving target and a number of people who identify themselves as Christians would exempt themselves from my criticism because they do not entirely agree with Collins's version of Christianity. However, I think most Christians believe in the divinity of Jesus, the Bible as a reliable moral guide, miracles, the efficacy of prayer, and an afterlife of reward or punishment. In short, they accept the existence of a supernatural reality. My scientific worldview does not accept and argues against these core Christian beliefs. Usually, the differences between Christians are about doctrinal details like the relative importance of faith and good works, original sin, the definition of sin, evolution, homosexuality, abortion, and so on.

It would be impossible in a single, readable book to discuss all the arguments used by various religious groups to justify their beliefs. My modest objective here is to consider the arguments and "evidence" used by one famous scientist to justify his belief in Jesus as the one true God. Since Collins's objective is to make Christianity and belief in Jesus as God sound, compatible, and plausible with scientific knowledge, especially evolution, I will not address other supernatural beings who may or may not be the subject of another religion's belief. But the same arguments are just as effective against other personal Gods such as Allah and Yahweh.

It is not necessary to read *The Language of God* to follow this critique because I believe I have fairly stated Collins's evidence and arguments. However, I encourage you to read the original because it is always enlightening to read an author in his own voice. And I do not cover Collins's explanation of genetic science nor his arguments in favor of the theory of evolution, with which I have no dispute.

Finally, I make no pretensions about resolving this long-standing debate between science and religion, but I do believe the "evidence" in Collins's book deserves critical examination. Like Collins, I hope my book will stimulate you to rethink and explore more deeply your own worldview, your own answers to the ultimate questions. You have the same capacity as Collins or me to analyze the world around you and evaluate the competing explanations. I am not dismissing Collins's book as a potpourri of periphrasis, which plagues the reader with a peripatetic plethora of pleonastic, platitudinous, pedantic pettifoggery, as I might if I intended to write an academic tome. Rather, I hope to communicate in general, common language of ordinary people who want to engage in investigating important questions. You do not need to be a scientist, philosopher, or theologian to understand this book. I welcome commentary on the arguments I set forth with a view toward extending rational dialogue on this important debate.

## NOTES

1. F. Collins, *The Language of God* (New York: Simon & Schuster, 2006).

2. K. Miller, *Finding Darwin's God* (New York: Harper Perennial, 2007); F. Ayala, *Darwin's Gift to Science and Religion* (Washington, DC: Joseph Henery Press, 2007); D. Falk, *Coming to Peace with Science* (Downers

Grove, IL: Intervarsity Press, 2004); J. Roughgarden, *Evolution and Christian Faith* (Washington, DC: Island Press, 2006).

3. Collins, *Language of God*, p. 2.

4. Harris Poll #60, October 16, 2003, http://www.harrisinteractive.com/harrispoll/index.asp?PD+409.

5. Collins, *Language of God*, p. 4.

6. E. Larson and L. Witham, "Scientists and Religion in America," *Scientific American* 281 (September 1999): 88–93.

7. Harris Poll #80, October 31, 2006, http://www.harrisinteractive.com/harrispoll/index.asp?DID=707 (accessed September 21, 2007). For discussion of historical trends, see G. Bishop, "The Polls—Trends: Americans' Belief in God," *Public Opinion Quarterly* 63 (1999): 431–34.

8. E. Larson and L. Witham, "Leading Scientists Still Reject God," *Nature* 394 (July 18, 1998): 313.

9. E. H. Ecklund and C. P. Scheitle, "Religion among Academic Scientists: Distinctions, Disciplines, and Demographics," *Social Problems* 54 (2007): 289–307.

10. http://pewforum.org/news/display.php?News/D=15039.

11. Gallup Poll, February 20, 2007, http://www.gallup.com/poll/26611/some-Americans-Reluctant-Vote-Mormon-72yearold-Presidential-Candidates.aspx; Association of Religion Data Archives, Opinion of Atheists, http://www.thearda/quickstats/qs_32.asp.

12. R. Dawkins, *The God Delusion* (New York: Bantam Press, 2006); S. Harris, *Letter to a Christian Nation* (New York: Knopf, 2006); D. Dennett, *Breaking the Spell: Religion as a Natural Phenomenon* (London: Viking, 2006); C. Hitchens, *God Is Not Great* (New York: Twelve, 2007).

13. http://www.templeton.org. For discussion of John Templeton Foundation and its interaction with scientists, see A. Saxon, "Sir John Templeton's Foundation and the New Trinitarianism," *Free Inquiry* (June/July 2007): 27–34; C. Holden, "Subjecting Belief to the Scientific Method," *Science* (News Focus) 284 (May 21, 1999): 1257 and comments, *Science* 284 (June 11, 1999): 1773; John Horgan, "The Templeton Foundation: A Skeptic's Take," *Chronicle of Higher Education*, April 7, 2006, http://www.edge.org/3rd_cultureshorgan06_index.html.

14. J. Lippard, "Trouble in Paradise: Answers in Genesis Splinters," *Reports of the National Center for Science Education* 26 (November/December 2006): 4.

15. J. Kirsch, *God against the Gods* (New York: Penguin Books, 2006).

16. R. Stark, *What Americans Really Believe* (Waco, TX: Baylor University Press, 2008).

17. G. Paul, "Is the Baylor Religion Study Reliable?" An analysis from the Council for Secular Humanism, http://www.secularhumanism.org/greg-paul-baylor.pdf (accessed February 9, 2009).

# *Chapter 1*

# FROM BELIEF TO ATHEISM

*There is no harder scientific fact in the world than the fact that belief can be produced in practically unlimited quantity and intensity, without observation or reasoning, and even in defiance of both, by the simple desire to believe founded on a strong interest in believing.*

—George Bernard Shaw

## A PERSONAL JOURNEY

Dr. Collins begins his book in a praiseworthy fashion by describing how he was raised and how specific people and places influenced his beliefs. Too many authors ignore the impact of these outside forces, making it difficult for readers to determine how the authors might be biased or if their work is an overreaction to some unusual occurrence in their past. Authors are humans, not disembodied, objective, and authoritative voices imparting wisdom via the printed page. To give the reader the same kind of personal insight, I begin with relevant material about my own upbringing.

In contrast to Dr. Collins's rural origins, I was raised in a city, San Francisco. I was born in 1930 in the depths of the Depression. My mother was an office worker with a high school education. She was a practicing Roman Catholic. My father was a truck driver with a third-grade education and did not practice any religion. I was the eldest of six children and, when I was young, we lived on government-funded programs of the New Deal.

Under Franklin D. Roosevelt's New Deal Program, the government served the interests of the common people, addressing their problems by enacting laws that established Federal Deposit Insurance to guard against failed banks; the Social Security Act to protect the old, disabled, and destitute; and the Works Projects Administration, which put the unemployed, including my Dad, back to work building bridges, tunnels, roads, sewers, water systems, libraries, parks, post offices, and airports.

I was born and baptized in the Holy Roman Catholic Church at the age of one month at St. James Church in San Francisco. Before I could walk, I accompanied my mother to mass every Sunday. And before I attended St. Agnes Catholic grammar school, I learned the classic prayers "Our Father" and "Hail Mary." The Nuns of the Presentation taught me my catechism, a series of questions and answers to be memorized: "Who made you?" "God made me." "Why did God make you?" "To know him, love him, and be happy with him in heaven." These were not simply memorized; they were emotionally accepted as true statements of fact. I made my first communion at around age seven.

After confessing my sins to the priest and doing penance to purify my immoral soul, I received what appeared to be a flat thin wafer but what I truly believed was miraculously the actual flesh of Christ. Out of respect, I would not chew the wafer, but allowed it to dissolve while I contemplated God's love and my commitment to follow his Gospel. I had my own set of rosary beads and during lapses in daily activities would finger the beads, mentally reciting "Our Fathers" and "Hail Marys." I used to get up early on Sunday to attend the 6:30 or 7 AM mass because I felt that many who came to midday mass were there to socialize and display their Sunday best and lacked the real spirit of humility and submission to divine instruction.

My beliefs were strengthened by the sacrament of Confirmation and reinforced by the Catholic ritual of making the Stations of the

Cross and giving up some innocent pleasure for Lent. I continued my study and practice of Catholicism at a Jesuit high school, St. Ignatius, and received a scholarship to the Jesuit University of San Francisco. St. Ignatius of Loyola, a Spanish soldier, established the Society of Jesus as the pope's "shock troops" in the Counterreformation. The order has always had a strong educational role and emphasized defense of the faith by reasoned argument rather than by passive acceptance. The Jesuits accepted, with St. Augustine and St. Thomas Aquinas, the principle that there was only one "truth," so there could be no philosophical or scientific truth that was irreconcilable with religious truth. Respect for truth has been one of the most important legacies of my Catholic education.

However, I became uncomfortable about some aspects of my religion. For example, the doctrine of original sin troubled me. It did not seem fair and just that a loving God would punish the descendants of evildoers like Adam and Eve. Why were their innocent offspring condemned to suffer unimaginable pain eternally if they were not baptized and had not accepted Jesus?

I certainly felt no guilt from my father's actions over which I had no control, nor would I want my children punished for what I might do. What about the millions of humans who never read the Bible, never heard of Jesus, and were never baptized? This was resolved in my adolescent mind by the Catholic doctrine that there are two kinds of baptism: baptism by water, as a formal ritual; and baptism by desire. If pagans wanted to be good, they would have accepted baptism if it were made available to them. If a pagan acted in good faith in accordance with his beliefs to do good and avoid evil, he was doing all that God expected of him and was "baptized" and part of the Catholic community. Since there was no salvation outside the Catholic Church, a doctrine recently confirmed by the Vatican Council II, a similar line of reasoning applies to Protestants. To my mind, a righteous Protestant was a secret Catholic by baptism of desire. A loving God had provided a get-out-of-hell pass. Likewise, the sentencing of stillborns and newborns who die prematurely to hell was something I found difficult. Again, I found an acceptable solution; they go to Limbo, a place without pain but also without the full rapturous joy of heaven. My ideal of a loving God was preserved.

Another concern was the definition of sin. If sin was any transgression of God's law, such as dishonoring or lying to your father and mother, and all persons dying in sin suffered the torments of hell, I was at serious risk of not getting to confession in time and ending up there. This worry was eased by the doctrine that there were two kinds of sins: mortal, which was very bad and deserved hell; and venial, which was not so bad and got you into Purgatory. Purgatory was a place of suffering, but not for eternity. When you had paid for your minor transgression, you got to go to heaven. And so, by rationalization and reinterpretation, my faith remained firm.

On graduation from St. Ignatius High School, I received a scholarship to the (all-male at the time) Jesuit University of San Francisco. The small class size and easy access to instructors promoted dialogue and debate. It also promoted critical thinking and was an excellent foundation for my education. My desire to help my fellow man and relieve suffering, and my fascination with living things led me to enroll in the premedical curriculum, which exposed me to science. This college experience stimulated my lifelong curiosity and enjoyment of acquiring knowledge. The result was that, beginning in college, I found inconsistencies between my religious beliefs and what I was learning in science. For a long time, I found acceptable rationalizations or stretched my credulity to reconcile the apparently incompatible truths of science and revelation.

My high school science courses also left me with the impression that physicians were glorified mechanics who merely applied scientific knowledge and discoveries, while medical researchers undertook the real challenges by preventing and curing disease and disability. I wanted to become one of these creative researchers and discover a cure for cancer or something that was equally likely to earn me a Nobel Prize in medicine. My family's financial circumstances dictated that my education could only continue if I received financial aid. So, after I graduated from USF, I applied for as many scholarships as I could and received one to the University of Southern California in 1952.

There was no single dramatic event that caused me to lose my faith. I just found it harder and harder to rationally accept the idea of a personal God, virgin birth, and resurrection, and to reconcile God with the existence of so much evil and the negative effect of religious excess. I stopped attending mass.

At this point in my intellectual odyssey, I was interested in viruses. I believed that studying these simplest forms of life could unlock the critical difference between living and nonliving things. So, I majored in bacteriology but soon came to realize that virology was not well represented at USC at that time. My next move was to apply for financial aid from the University of California at Berkeley, where a world-class virus laboratory had been established, directed by Nobel Prize winner Wendell Stanley. In 1953, I was awarded a paid research assistant position and transferred to Berkeley.

In the course of my studies and work at the virus laboratory, I became aware of the deficiencies of academic science as a social institution. There were many instructors or assistant professors who worked for low salaries and were constantly in competition with one another for tenured appointments. I did not find my idealized concept of university scientists engaged in a collaborative, cooperative search for objective truth and knowledge.

My faculty advisor was a good example of the problem, who, having worked at several universities, had to then move on when he failed to get tenure. His plight reactivated my interest in a medical degree. Doctors have direct contact with people who can benefit from their services, can do research just as well as academic scientists, and have more independence, better security, and higher salaries.

So, in 1955, I abandoned my quest for a PhD and applied for scholarships to the major California medical schools. I was accepted at USC and Stanford, but I could not afford the tuition. I was also accepted by University of California at Los Angeles. As a California resident, tuition was affordable if I could work part-time and during summer vacations. I accepted and began my medical training there.

Collins felt challenged by the faith and religiosity of his medical school patients who asked him questions he had not previously tried to answer. In contrast, I had long recognized the importance of answering these questions and was not disturbed by other people's faith. The personal challenge I experienced was the suffering of good, innocent people, especially children, and the unanswered question, "Why does an all-powerful God permit this?" What kind of a creator would produce such an imperfect creation? In contrast to Collins, who experienced life-changing illumination while reading C. S. Lewis's *Mere Christianity*, there was no one book that lifted my veil of ignorance and provided a definitive answer. I had read many books, including the Bible, on both sides of the God issue.

My experiences as a medical student, and probably my natural inclinations, led me to believe that I would find most satisfaction as a pediatrician. Children have more curable conditions and have their whole lives before them. They are still capable of trust and unconditional love and are generally positive and hopeful. So, upon graduation, I applied for a straight internship in pediatrics at UCLA. This decision was prompted, in part, by a dramatic case I covered during my fourth year of medical school.

I had made the mistake of lingering around the hospital after all my assignments were completed. Suddenly, the chief resident directed me to report to the operating room. There I found myself one of a large team of doctors and nurses working on an elderly man whose major artery, the aorta, had ruptured. At the time, UCLA had some of the leading cardiac surgeons, who were struggling in an abdomen full of blood to find the tear, stop the bleeding, and sew in a Dacron graft. The bleeding was massive. They transfused unit after unit of blood. Over twenty units were used and they were coming from the blood bank refrigerator without warming to room temperature, so the patient's temperature was dropping dangerously.

I stood there for what seemed like hours holding back the liver with an instrument so the surgeons could do their work. Not due to a miraculous intervention of God, but due to the skill of the surgical team, the bleeding was stopped, the new tubing was sewn into place, and the patient was moved—in critical condition, but very much alive—to the surgical ward.

I sat up in the room with this patient all night, fitfully sleeping in the chair and observing his vital signs. I reviewed his medical record and noted that he was over seventy years old and had no living family. He had failing eyesight and hearing, severe arthritis, and other chronic conditions requiring medication. We had saved his life, but what quality of life would he have? How many more years would he live? Early the following morning he became delirious, thrashing around. Despite the efforts of the emergency team, he disconnected his intravenous infusions and quickly went into shock. We tried to restart his heart and gave him infusions, but he died. This was not an emotionally positive experience. What did we have to show for all our efforts? How much good had we done? Certainly this man deserved this heroic effort, and these questions should not be misinterpreted as a criticism of the staff for providing care. I am glad that there are physicians and nurses who find this kind of activity rewarding and challenging. All I am saying is that I learned that I would rather spend my time in a different area of medicine.

In contrast, a lethargic and feverish young child was brought to the emergency room. Detecting pain when the spine was stretched, I did a spinal tap and obtained cloudy, not clear, spinal fluid. I started antibiotics immediately to treat meningitis. When I discharged this child some days later, it was with a very positive sense of accomplishment. I was personally rewarded knowing I had probably saved both his brain and his life. A lifetime of opportunities was preserved.

During my third year in medical school, I met and married a grade school teacher. After I completed my internship, I was accepted as a resident at Children's Hospital of the East Bay, and we moved back to the San Francisco Bay Area. Following my residence, I spent several years at CHEB as staff fellow doing research on iron poisoning and metabolic disorders. Then, in 1965, I joined the California Department of Public Health to initiate a Hereditary Defects Unit. I took the position because I knew I could influence the health and welfare of literally millions of newborns and their families.

I served for forty years, protecting and promoting the public health in California in the areas of maternal and child health and genetic disease. Although I retired as chief of the Genetic Disease Branch for the California Department of Public Health on December 31, 2005, I continue to educate myself on advances in scientific knowledge and continue my studies of major religions and the works of philosophers. With this brief background, you now understand the process by which I arrived at my beliefs about God. I now hope to engage you in a dialogue and analysis of what Collins feels is rational, logical, and even scientific evidence for the existence of a personal God.

*Chapter 2*

# EVIDENCE AND RULES OF ARGUMENT

*Facts are stubborn things, and whatever may be our wishes, our inclinations, or the dictates of our passions, they cannot alter the state of facts and evidence.*

—John Adams

## REFERENCES AND SOURCES

To determine if Collins's arguments are valid and his evidence sufficient to support his contention that he has reconciled science to his faith, you must read his references and not ignore any of his supporting material. While Collins claims to provide the evidence that persuaded him to accept Christianity, for some reason he rarely cites books or scientific articles to support or challenge his opinions and conclusions. His conversational style makes for easy reading, but at the cost of omitting and underanalyzing crucial evidence. He uses words such as "spiritual," "faith," "belief," or "evidence" that he only vaguely or inconsistently defines, making his argument hard to follow. I have included notes at the end of each chapter referencing studies and books that bear on the quality and quantity of evidence I present. When I quote from Collins's book I include page numbers. All quotations from the Bible are from the authorized King James Version.

## BELIEF, KNOWLEDGE, AND FAITH

In order to analyze the arguments and evidence Collins offers in his book, it is necessary to define the basic concepts of belief, knowledge, and faith. I cannot use Collins's definitions because he doesn't provide them. The definitions I use are not invented by me but are widely used in philosophy, science, and theology.

There are some things about the universe we know, some things we believe, and some things we do not believe. There are a few things called *facts* that we know as true all the time and everywhere. These things are self-evident; that is, they require no evidence and do not depend on anything else to be true. The knowledge that you exist, that a real universe of things outside of you exists, that ten is greater than two, that all bachelors are unmarried, are accepted without proof. The principles of cause and effect and the uniformity of nature are not accepted on faith, as some philosophic critics of science would have us believe. They are self-evident from our daily experience and require no proof.

This is one kind of knowledge, self-evident knowledge, but if you think about it, most things you know about the universe are really just beliefs justified by a sufficient quantity and quality of evidence. Belief is a mental state of the believer that accepts some statement or claim about the nature of reality as to some degree representative of actual reality. A belief is true to the extent that it actually does represent the real situation in reality and false if it does not reflect reality.

You believe that day follows night and that a pan of water on a hot burner will boil because you have the evidence of your own perceptions and experience reinforced by the reports of other qualified observers. You may feel that you know this is a fact that is 100 percent certain, but actually there is a very small probability that day won't follow night or the water in the pan would freeze. For all practical purposes we use the term *certain* to mean a very high degree of probability, but most knowledge is probabilistic. Knowledge—that is, not self-evident knowledge—is defined as belief that is well justified by a sufficient quantity and quality of evidence. Since we are defining knowledge as justified belief, and evidence for a belief is never complete, even what we think we know today may have to be discarded or modified in the future if new evidence becomes available. True knowledge is true to the extent that it corresponds with objective reality. Science tries to gain knowledge about the real world but can never attain a complete and absolutely true understanding of the universe because the universe is constantly evolving and producing new evidence and phenomena. However, science is the best knowledge we have and as time goes on is making better and better approximations of the state of the real world.

When we believe something is probably true but regard the evidence supporting it to be weak or insufficient, we call this an *opinion.* It is my opinion that life exists elsewhere in the universe, but I don't have sufficient evidence to be sure. A belief or opinion can be called *rational* as long as it is not self-contradictory and we have some minimal evidence to support it—but this does not mean that it is *true*. Can you have rational reasons for believing something that is not true? Yes, in certain situations this is possible. For example, assume we have a mutual friend and we say good-bye to him at the pier in San Francisco. We watch him sail out to sea to fish for the day. The next morning we find out he has not returned. You would not be irrational to believe him to have drowned at sea. You might cite evidence to support this belief. He has never remained at sea for more than a few hours in the past. He has night blindness and would have difficulty maneuvering in the dark. There are treacherous currents and tides in the area. It would be reasonable for you to believe he was dead, but your belief could only be as strong as your evidence. The probability of your belief being true is dependent on the quality and quantity of evidence. I might rationally believe that our friend is lost but not dead and I might also cite evidence to support my belief. Given what we know, both of us have some rational reasons for our beliefs. Over time more evidence would be added to strengthen or weaken our respective beliefs about whether our friend is dead. Only one of us would be proven by the accumulation of good evidence to have had a true belief, while the other would have held a false belief. We now can say we know what happened to our friend. At that point failure to accept the evidence becomes irrational. If we both possess the same strong, sufficient evidence in support of a belief and we still fail to accept the belief as true knowledge, then one of us can be properly described as irrational.

Belief without supporting evidence, or in the face of evidence that proves it is false, is blind faith. It is not knowledge in any sense, however defined. When someone expresses such a belief without any justification, you as a reasonable person are under no obligation to accept it as true. For example, it is not necessary for you to believe the unsupported claim that Muhammad ascended to paradise from Jerusalem or that herds of English-speaking unicorns live in Southern Africa.

Theists do not want to accept this common definition of faith. I don't know what they would call holding a belief in the absence of evidence. Perhaps they would call it superstition. While we are considering the question of whether Collins has provided reasonable evidence for what he calls his faith, we have to be clear whether we are using "faith" as a noun to describe a set of beliefs or as a verb describing the process of using trust and intuition to accept a belief or set of beliefs. Paul in his letter to Hebrews gives us a biblical definition of faith: "Now faith is the substance of things hoped for, the evidence of things unseen." This is an admission that faith is hoping for things. The wish for something better. Paul bases this faith on "evidence" beyond the senses (seeing). This is a misuse of the word *evidence*. It is hoping something is true without "seeing" any evidence that it is. How do we collect evidence of things unseen? All evidence that science uses is seen either directly or indirectly. Quarks and dark matter, for instance, are accepted as part of reality based on their observed effects. Quantum theory is tested by measuring observed results.

You do not have *faith* in the truth of things that you know by evidence and experience to be true. No one standing in the rain asks you to accept on his or her word alone that it is raining. You both *know* it is raining. But if the person asked you to believe that the rain would cause your sex to change on his word alone, you could only accept it on faith. Scientists do not want to accept anything on faith as we have defined it, that is, something with no evidence or reason to support it or even evidence that it is untrue. If a famous scientist like Francis Collins announced that he had by a simple and inexpensive process cloned a human being, some scientists might accept his claim tentatively on faith out of respect for his reputation. But, they would expect to see strong evidence and replication by other scientists before they would accept it as a fact they know.

## WHAT IS VALID EVIDENCE?

Let's consider whether you would find the evidence in *The Language of God* sufficient to accept, first, a belief that there exists such a being as God, and second, that this being is or was manifested as Jesus. *Evidence* is any relevant information that contributes to the probability that a statement is true. What kinds of evidence would you look for and how much evidence would you need? Renowned philosophy professor James Hall provides an analysis of the kinds of evidence potentially available, which I have summarized in Table 2.1.[1]

Since, as Collins admits, "rational argument can never conclusively prove the existence of (a personal) [*sic*] God,"[2] his objective is to make the belief that Jesus was God "plausible." *Plausible* does not mean *probable*. If we accept the definition that knowledge is true belief justified by a sufficient amount of the right kind of evidence and faith is belief without sufficient evidence, we need to determine what counts as evidence for or against the theistic position. There are different kinds of evidence. There is *factual evidence* directly confirmed by personal and community experience, for example, all humans die. There is *logical evidence*, for example, if I tell you I have a cat or snake

TABLE 2.1

| TYPE | DESCRIPTION AND COMMENTS |
|---|---|
| Reason | Deduction: from a universal, self-evident, true statement to a specific case or situation or observation. |
| Induction | From a large number of cases or observations to form a general rule that is true. |
| Analogical | From cases or observations that share common factors and derive a common cause or explanation that may be true. |
| Appeal to Authority | Reliance on truthful testimony of qualified experts in the area under investigation, that is, secondhand experience. |
| Intuition | Immediate insight or impression of some sort of possible explanation or relationship of observations. Not acceptable by science as evidence without further supporting evidence. |
| Revelation | Only used to support "religious truth"—actually a special case of appeal to authority of god or witnesses. Not acceptable by science. |

in a box, you know that if it is a cat it cannot be a snake and vice versa. There is *definitional evidence*, for example, for any and all triangles in a two-dimensional plane the internal angles add up to 180 degrees in Euclidean geometry. There is evidence that relies on an accepted set of *values*, for example, torture is always wrong. In making arguments for or against theism, you can use direct or indirect validated experience, for example, I know Rome exists directly because I've been there, or I know Tokyo exists indirectly because the accumulated reports of its existence by unbiased, competent observers are convincing. Experience needs to be validated by controlling variables that affect the result using accurate instruments and by eliminating bias on the part of the observer.

Reason is the use of our minds to evaluate the relationships between the facts of experience so we can create a consistent, integrated understanding of how the universe operates. Reason seeks true knowledge. The process of reasoning is either *deductive*, going from a universal or previously proven truth to apply it to a particular true example, or *inductive*, going from a number of individually true examples to derive a universally true general conclusion. These valid and true conclusions can be used as evidence in building an argument. As an aid to determining the truth of an argument, Aristotle developed a form of argument called the *syllogism*. A syllogism consists of statements called *premises* that provide information about the subject of the argument. Rational proof consists of a series of true statements or premises, which, if the rules of logic are correctly used, require that the conclusion that follows from these premises be true. The first principle of logic and science is the principle of *noncontradiction*, namely, it is impossible for something both to be and not to be at the same time and in the same respect. An example of contradiction is the statement that God is one person and is also three persons at the same time and in the same respect. Another principle is called the principle of the *excluded middle*, which states that at any point in time a thing either exists or it does not exist. There is no middle state. For example, a personal God must either exist or not exist—there is no alternate. These are called principles of logic because they are self-evidently true and require no proof. Reason can be used to develop information, that is, evidence, useful in supporting an argument. We also reason by analogy when we apply information about the characteristic of something well known to something different and less well known but similar. In order for the analogy to provide useful information or to function as evidence supporting an argument, the degree of similarity between the things or events used must be high in all significant respects. The comparison does not lead to necessarily true information. Intuition, a spontaneous insight, hunches, individual anecdotes, single examples, unverified reports of experience, and the like are not really evidence in the sense I am using the word.

Is it rational to believe something on the basis of testimony? Yes, it is. We all accept testimony as the basis of our belief but we do not accept it unconditionally. We generally require that the testimony be from unbiased individuals. We expect they would be competent observers who could distinguish reality from illusion or delusion. There are some things that are so incredible that we would not believe them on testimony alone. If the truth of the belief is really important and would affect your whole view of life, then you should require more than testimony, however sincere. Accepting second- and thirdhand accounts by biased witnesses of very improbable events as evidence for your worldview is irrational. Many studies have shown the weakness of even firsthand witness testimony.[3]

Can divine revelation be used as evidence? If it could be convincingly and independently established that the source of the information was in fact God, it would qualify as evidence. However, before you could appeal to revelation, you would need to establish the existence of the deity. Revelation depends on prior proof of God. Assuming you succeeded in proving God's existence, you would then need to prove that the alleged revelation did in fact come from this God.

Evidence—reasons justifying belief—varies in quality. Eyewitness testimony is usually better than second- or thirdhand accounts. Many eyewitnesses are better than just one. Objective, unbiased witnesses are better than biased witnesses. Expert testimony is better that nonexpert. A broad consensus of experts is better than a single expert. Evidence that can be repeatedly tested is better than one-time occurrences. Evidence that is consistent with previously accepted truths is better than inconsistent evidence that creates more questions than it answers.

## BURDEN OF PROOF

The greatest philosophers and theologians throughout history have been unable to produce a valid logical proof of the existence of the supernatural being called God. A valid proof could only be an argument from clearly true or self-evident premises that lead to a necessarily true conclusion. So it is unfair to require Collins to bear the burden of proving his God exists. Collins acknowledges that such a proof does not exist, and so his more modest aim is to show that his faith is not contradictory or incompatible with the knowledge developed by science and in fact is reasonable in the sense that it is the best explanation of the evidence. Collins has the burden to prove not that his version of God exists but that it is rational to have faith that such a God exists. My task is not to prove a negative, namely, that God does not exist. If you believe a herd of invisible, English-speaking flying serpents are living in Antarctica, it is not my task to disprove it but only to ask you for good reasons, namely, evidence to accept this belief. I do not have to prove that science gives us the best knowledge of the natural world. Collins accepts this common belief. Collins has to produce evidence that, in addition to the natural universe, shows there is a supernatural reality containing his God. I am not asking Collins or you to believe anything radical or new. We all live in the natural world. But he is hypothesizing the existence of a supernatural reality, which is not self-evident and is very controversial.

The evidence assembled must be used in proper logical form to preserve the truth. The arguments must be properly valid. Circular arguments like "the Bible is the word of God because the Bible says it is it the word of God" are not valid. The conclusion of an argument must necessarily follow from its premises: "all priests drink alcohol, the pope is a priest, and therefore the pope drinks alcohol" is a valid but not necessarily true argument, while "all priests drink alcohol, Ann Coulter drinks alcohol, and therefore Ann Coulter is a priest" is an invalid and inconclusive argument.

Finally, after you have collected acceptable evidence, how much evidence do you need to achieve justification for your conclusion or argument? Since absolute certainty is a rare outcome, I suggest we use the legal standard, namely, sufficient evidence to accept the argument beyond a reasonable doubt and with a very high probability of being true. We should accept the best explanation consistent with all the evidence. In our quest for reasonable answers, we can consider all kinds of evidence as Collins and other believers use them, but we also must consider how much evidence is needed and how strong it must be. The belief that "Beethoven was deaf" needs to be supported by evidence. But the truth of this belief is not of major importance, so we could accept it with minimal evidence. However, the belief that we live in a universe in which a supernatural being controls all human experience and desires a fellowship with its creatures would seem to need a lot of very good evidence, given its importance.

Collins bases his belief in a supernatural deity on four arguments. First, he questions whether you can be good without God providing a universal moral law. Second, he contends that the only explanation of the big bang and the cosmological constants of physics is an intelligent creator God. Third, he argues that only God provides a meaning and purpose for our lives. Finally, the universe is so beautiful and well ordered that it must be the product of a divine creator. After he makes these arguments for God's existence, he then argues this God became incarnate in the person of Jesus Christ. His only evidence for this claim is the Bible. Let us examine in detail the evidence for each of these claims using the rules of argument discussed above.

## NOTES

1. J. H. Hall, *Philosophy of Religion*, Course Guidebook, 2003, Teaching Company Lecture, http://www.teach12.com.

2. F. Collins, *The Language of God* (New York: Simon & Schuster, 2006), p. 164.

3. D. F. Ross, J. O. Read, and M. P. Toglia, eds., *Adult Eyewitness Testimony: Current Trends and Developments* (New York: Cambridge University Press, 2007).

## *Chapter 3*

# THE WAR OF WORLDVIEWS

*Only when truth is no longer confused with mythology; only when transcendent rewards or punishments are no longer postulated in an after-life; only when people judge other people by their actions rather than by their beliefs, or lack of beliefs; only then might it be possible for the concept of absolute evil to wither away, to be replaced by a new ethic of kindness and tolerance. Will it happen, sooner or later? Will it ever happen? What do you think?*

—Barbara G. Walker, *Freethought Today*, September 2007

Before making his case for God, Collins addresses the general concept of the war between the scientific and religious worldviews. Collins acknowledges that most scientists are not Christians and have difficulty accepting the reality of a personal God. Barriers to belief about the existence of a personal God, according to Collins, can be based on "perceived conflicts of the claims of religious belief with scientific observations" or on "philosophical issues."[1] Collins identifies four "particularly vexing" barriers to belief that he addresses in the second chapter of his book. These barriers are (1) wish fulfillment, (2) harms done by religion, (3) the existence of evil, and (4) miracles.

## WISH FULFILLMENT

The first barrier Collins seeks to knock down is Sigmund Freud's characterization of belief in God as "wish fulfillment." Freud explains theism as a wish for a perfect father in place of our imperfect human fathers and "a universal but groundless human longing for something outside ourselves to give meaning to a meaningless life and to take away the sting of death."[2] After all, aren't we all children of God when we call on "Our Father, who art in heaven"?

Collins refutes Freud's claim using three arguments: first, wishing for something has no bearing on whether or not that something exists. I agree. Second, he agrees with C. S. Lewis's assertion that "creatures are not born with desires unless satisfaction for those desires exists."[3] Not true. Men desire immortality, absolute certain knowledge, infinitely satisfying love, invisibility, communication with the dead, and time travel. All of these are real but unattainable desires.

Finally, Collins rejects the concept of wish fulfillment because "it does not accord with the character of the God of the major religions."[4] God's character as "described in the Bible" is in Collins's view an authoritarian enforcer of a strict moral law and not the "benevolent, coddling," and indulgent father figure we would wish for. Collins has not yet established that the Bible is a reliable source of information about God, but many would argue that the character of Jesus is just the kind of supportive, fair, loving, and forgiving father most people would wish for. Collins assumes facts not in evidence, namely, the truth and accuracy of the Bible. This is a major issue, which I will address in chapter 6. Moreover, people reading the Bible tend to come up with their own view of the personality and character of God, views that are different from Collins's. The God they find in the Bible may be exactly the kind of God they wish for. Many Christians and non-Christian theists do not accept the Bible as the absolutely true word of God. They are free to wish for and construct a God that would satisfy their particular personal needs.

One consequence of the evolution of the human brain is the

development of curiosity about the world and the need for satisfying explanations of the meaning of things and events, especially death. This desire to know the certain, final, eternally true purpose and meaning of our universe and our lives creates a longing that can never be completely satisfied by science or reason. Atheists generally deny such an external purpose exists and find meaning in our interpersonal relationships. Theists believe that this desire can only be satisfied by accepting an imaginary something beyond the natural world, namely, God. In this sense, wish fulfillment partially explains the persistence of theism in the modern scientific age. Wish fulfillment as a mental exercise does not prove nor disprove the existence of God but establishes a common emotional bias when examining the evidence.

The concept of wish fulfillment was popularized by Sigmund Freud as part of his theory on the interpretation of dreams. Collins confuses philosophy, the systematic analysis and critical examination of fundamental problems concerning the nature and operation of the universe and man's place in it, with psychiatry, a branch of medicine that deals with cognitive or emotional disorders. A number of philosophies are based on the concept that the natural world is less than real because it is perceived through our imperfect senses. The result of our misperception is suffering and/or lack of knowledge. The material world we perceive with our senses belies the real, ideal world of philosophers like Plato, Kant, Berkeley, and Schleiermacher that lies beyond our day-to-day world of experience. Buddhism and its emphasis on meditation and separation from the natural world is an example of another response to imperfection and suffering. But these are not what Collins has in mind, so they will not be considered.

There is little that can properly be described as systematic philosophy in Freud's writings, or for that matter, in the writing of C. S. Lewis. Again, Collins seems persuaded by Lewis's imaginative writing, in this case his book *Surprised by Joy*.[5] This is Lewis's autobiography in which he relates events and experiences during which he unexpectedly felt brief but deep joy and satisfaction. Lewis speculated that this "joy" was the result of the temporary fulfillment of some deep, unconscious desire. What are we longing for? Why, God, of course. The things that triggered Lewis's joyful epiphanies were, in keeping with his poetic nature, a toy garden given to him by his brother; reading Beatrix Potter's children's story *Squirrel Nutkin*; phrases in books or poems; and so on. Collins recounts various incidents during which he felt a "special kind of joy associated with flashes of insight."[6] Many, if not most, scientists have experienced this extremely pleasurable "A-ha!" moment. Most of us have had similar experiences in response to significant pleasurable events in our lives. I know I have felt this kind of joy. But need we, can we, interpret this transient joy as anything more than a complex and powerful surge of the emotional centers in our brains? This is not, as Collins claims, "an experience that defies a completely naturalistic explanation."[7] Collins denies scientific facts when he claims it cannot be explained in terms of "some combination of neurotransmitters landing on precisely the right receptors."[8] The same emotion could be naturally created by stimulation of the right receptors in the brain during brain surgery. Brain-damaged humans who lack this neurological mechanism do not experience this emotion. Clearly this joy can be produced by events that cannot be described as supernatural or even related to anything divine. Again Collins cites this joy as evidence of God's presence and action on humankind, dismissing or ignoring the natural explanation and leaving only the "God of the gaps" as his intellectually satisfying explanation. It is a real emotional experience and, like most emotional experiences, it is difficult to describe in words but is nonetheless entirely natural.

Collins points out that as individual humans we feel puny and limited in both what we want to accomplish and what we actually accomplish. We feel a need to belong to "something bigger than ourselves." This is why we form social groups around political causes, personalities, sports teams, and religious communities. These all, in various proportions, contribute to giving our lives meaning and purpose. The fact that we humans as intelligent social animals create and believe in things and causes and concepts bigger than ourselves is no proof that these things, causes, and concepts exist in the real world.

## WILL WISHING MAKE IT SO?

Collins asks, "Finally, in simple logical terms, if one allows the possibility that God is something humans might wish for, does that rule out the possibility that God is real? Absolutely not."[9] But does it rule out the possibility that God is unreal? Absolutely not. Clearly we can desire, long for, wish for a variety of objects and events both real and imaginary. You can wish for a life partner who is always loving, always loyal, and always completely satisfying and supportive of your every need. Does that wish, however real, guarantee that such an individual exists? True, you may find a partner who has many of these characteristics some of the time. But it is very unlikely, though not totally impossible, that you will find someone who meets all of the criteria all of the time.

The fact that humans wish for God cannot logically be used to prove that God exists or even to prove that God's existence is possible. This argument, which assumes all human desires can ultimately be fulfilled and, therefore God must exist to satisfy them, is flawed logic.

To analyze this longing for "something greater than ourselves" in naturalistic rather than supernatural terms requires science, rationality, and logic, not the poetry and speculation provided by Collins. The human brain is genetically programmed to operate in specific ways. We are not only conscious but also self-conscious, an ability only shared in a small degree with chimpanzees and possibly dolphins. Human self-consciousness includes the appreciation and acceptance that all natural life, including our own, ends. Our complex brain can desire and imagine an afterlife. The natural world is full of unfulfilled desires, such as the elimination of pain, evil, and injustice. The life after death is the only place we could imagine where all such desires could be satisfied.

Collins and Lewis argue that because we long for love, freedom from pain, justice, and immortality, these longings must be satisfied in an afterlife for all eternity. But this states as a logical necessity what is not logically necessary. It is conceivable that our longing for perfect love, certain knowledge, immortality, perfect justice, and freedom from life's imperfections only end with death, never to be satisfied. Collins's and Lewis's contention that all desires are satisfied by some real object is a proposition requiring a sufficient quantity and quality of evidence—in other words, proof.

Since there is plenty of evidence that our desires are not permanently and completely satisfied in this life, the only possibility is a life that transcends death. Where is the evidence, scientific or otherwise, for an afterlife? Where is the evidence that our desires are fulfilled there? Collins and Lewis fail to supply any evidence at all. In the last 50,000 years over 150 billion humans have died. We know that our physical bodies decay and there is no well-investigated instance of any human returning from the afterlife to contact the living or to resume life in human form. (Near-death experiences and biblical resurrections will be analyzed later.) The belief in another life after death is not universal. When asked about the afterlife, 74 percent of Americans believe in heaven but only 59 percent believe in hell.[10] Is this wishful thinking? How sure are you that your beloved deceased were really sinless in the eyes of God and will join you in the afterlife? Without their faces, bodies, voices, and physical mannerisms, how will you recognize them? If you think they will have new bodies, what will these bodies look like? What age would they be? Will they wear glasses? Will they have the conditions they had in life such as deafness, blindness, cancer, lacerations, wounds, and amputations? Collins does not present any evidence for the afterlife, nor does he speculate on its nature. However, since he does believe in the supernatural, I will discuss the evidence for the supernatural in chapter 8.

## CAN SCIENCE EXPLAIN THE NEED FOR RELIGION?

Collins contends that all humans feel incomplete without God. They need to have a connection with something supernatural; they have a

"God-shaped vacuum" that needs to be filled. Some humans, such as Buddhists and Taoists, do not seem to have this vacuum. Whatever the nature of this longing, I assert that it is biologically based and culturally conditioned. For a more complete scientific explanation of the persistence of religion and its biological basis, I refer the reader to the excellent book by physician/geneticist David E. Comings, *Did Man Create God? Is Your Spiritual Brain at Peace with Your Thinking Brain?*[11] I will only briefly address Collins's failure to acknowledge this body of evidence. Collins discounts the book *The God Gene*, by respected geneticist D. C. Hamer, who says there is a gene that influences a sense of transcendence.[12] Collins discredits Hamer because his research, while published in book form, has not been subjected to scientific publication and review. Collins also fails to mention an earlier report by David Comings that the dopamine D4 receptor gene is associated with the tendency to be spiritual and accepting of the transcendental.[13] Collins knows that there is little support or public or private funding available to research the genetic basis of religious experiences. But what will Collins's response be if and when future work confirms that constellations of genes determine our susceptibility to religious feelings?

It is already a scientific fact that deliberately stimulating the temporoparietal region of the brain by various natural actions produces a mystical religious experience, a feeling of being "one with the universe."[14] V. S. Ramachandran at UCLA demonstrated that patients with temporal-lobe epilepsy have heightened religious response.[15] Andrew Newberg used a PET Scan (single photon emission computed tomograph) to examine the brains of meditating Buddhist monks.[16] A PET device allows scientists to visualize the blood flowing through the different parts of the brain by using radioactive particles. He found decreased brain activity in the posterior superior parietal lobe, an area that helps us locate ourselves in three dimensions and separate ourselves from the world outside. Newberg believes that without the parietal lobe, the concept of god or God would not exist. No scientist or rational person would deny that while god or God might exist elsewhere, god or God also has to exist as a mental state in the brain. But this is where unicorns, fairies, and dragons also reside, and we find no evidence of their existence elsewhere.

The human brain can clearly construct imaginary beings that have no reality in the universe. You know that you have inherited the ability to create meaning and purpose, and you sometimes use it to find meaning and purpose when there is none. Sometimes the attractive stranger selects the seat next to you because it is nearest to the emergency exit and not because the person wants to make your acquaintance. Clearly the ability to determine meaning and purpose based on the signs and actions of other animals and humans has survival value. Movement in the tall grass ahead, alarm calls, growling, or baring the teeth—ignoring or misinterpreting any of these could be fatal. The god-shaped vacuum that developed in many brains is due in part to the extension of these pattern-finding, meaning-making survival skills to the more difficult task of find meaning in death itself. In most Western societies, the psychological longing for transcendental meaning and ultimate meaning and purpose of life itself is strong. While this is only a partial explanation of the persistence of religion, the desire for a loving father and for immortality is based on wish fulfillment.

## HARMS OF RELIGION

For Collins, the next vexing issue, barring belief by scientists, is the assertion that great harm is done in the name of religion. These harms are well documented in books by Hitchens, Dawkins, and Harris, among others. The fact that these authors express such moral outrage at religion's harmful effects does establish that atheists have a set of moral values based on humanistic reason, and they refute the charge that there can be no basis for morality without God.

Collins offers "two answers to this dilemma." First, he says, "many wonderful things have also been done in the name of religion."[17] While true, this "the good justifies evil" argument could also be applied to the Islamo-fascists, Nazis, Marxists, and the imperial colonizers of the world. These harms could be said to be the result of misinterpretation or misapplication of the ideas and ideals of the particular "ism." For example, real Muslims don't kill unbelievers; they give to the poor. Real Marxism promotes equality and does not involve gulags, and real colonizers educate and improve their subjects without exploiting them. Harms done by religions or atheistic political ideologies cannot be justified or excused by doing any amount of good.

Collins's second response to the idea that harm is done in the name of religion is his assertion that any such harm was done outside the tenets of true religious teaching. But which of the many interpretations of religious teachings, including the Bible, is the "true" interpretation? Does the true interpretation depend on when and where and who does the interpreting? Who is to say whether Amish, Mormons, Unitarian Universalists, and Christian Scientists are "real" Christians? Do you feel Francis Collins, Pat Robinson, Jim Jones, David Koresh, Dick Cheney, and Barack Obama are "real" Christians?

To use Collins's example, what is Christianity's teaching on witches? Do witches exist? What about Wicca? Does this qualify as witchcraft? Do we justify the arrest, punishment, and execution of anyone believed to be a witch? Is it Christian to execute or imprison physicians who terminate unwanted pregnancies? Are we justified in assassinating them? Is the harm done by opposing embryonic stem-cell research or contraception or gay marriage and adoption balanced by a greater good?

A variety of this defense is that mainstream Christianity is a source of good, and that the evils associated with it are the work of extremists and fundamentalists. There is broad agreement that the evil acts of fundamentalists should be condemned. But what about the goodness of mainstream Christianity? I do not know whether you would classify the psychological harm done to many children, in terms of the fear, anxiety, and loss of self-confidence and self-esteem that is imposed by the message of unavoidable guilt and eternal punishment imparted by believing or manipulative parents or ministers, as harm done by mainstream Christians. What about the physical harm inflicted on the kids? Federal law, the Child Abuse Prevention and Treatment Act of 2003, contains the following provision: "Nothing in this Act shall be construed as establishing a federal requirement that a parent or legal guardian provide a child with any medical service or treatment against the religious beliefs of the parent or legal guardian." Most state laws, including those in my home state of California, have religious exemptions from medical treatment, including mandatory public health programs like immunizations and newborn screening for preventable disorders. "Between 1975 and 1995 at least 172 children died in the United States because their parents refused medical treatment on religious grounds. One hundred and forty of those children died from conditions which medical science had a 90% track record of curing."[18] There are no statistics on how many more children were permanently handicapped, suffered unnecessarily, or died prematurely as the result of refusal to use modern medical treatments. In my opinion, the law should protect these children from preventable harm, notwithstanding the religious concerns of parents and guardians, by removing the need for parental consent for clearly necessary and clearly effective medical interventions. The religious rights of the parents are not violated since they are not personally transgressing the tenets of their religion. Nothing prevents them from praying or using additional religious rituals to assist in the child's recovery. Do you think that the minor who is not competent to consent to medical treatment is competent to consent to be bound by other people's decisions on his medical care based on other people's religious opinions? There is no assurance that when the child reaches the legal age of consent that the child will agree with his parents' or guardians' religious beliefs. The state should act in the best interests of the child based on the best information available, even if one or both parents consider the child's spiritual life more important than the child's health or survival. These religious exemption laws are immoral and do more harm to innocent children than good. Since only a small minority of parents or guardians uses these exemptions, why don't federal and state legislatures repeal them? Because these legislators are, with few exceptions, mainstream Christians whose private religious beliefs prevent them from acting in the interest of good public policy.

Mainstream religion teaches that the pope, bishops, rabbis, ministers, imams, and mullahs should be male. Mainstream religion has historically failed to allow women to participate fully in either civil or religious societies.[19] Science and politics have done more to include women and their concerns, even though it has been on a limited basis.

Mainstream religion simply reflects a majority opinion at a given point in time. One person's mainstream teaching is another person's heresy. In the past, most mainstream religions have opposed divorce, contraception, premarital sex, women's suffrage, alcohol, business on the Sabbath, gambling, homosexuality, and abortion. At different points in history, mainstream religion has supported the divine right of kings, slavery, censorship of speech and writing, and the execution of heretics and witches. So, contrary to Collins's claims, some of the real harm done by religions is done in "the name of religion" and does not "fly in the face of principles" of religion.

## THE PROBLEM OF EVIL

Collins's first two vexations are real, but minor, barriers to scientists' and rational thinkers' acceptance of God's existence. The third vexation, however, is an insurmountable barrier. This is the question of how a loving God that permits innocent humans to suffer can exist. Theologians and philosophers have given this question considerable attention, yet none have proposed an answer that is generally accepted as logically valid or empirically true. In fact, a bibliography covering the period from 1960 to 1990 listed 4,200 philosophical or theological writings on this irresolvable contradiction.[20]

The strong argument against the existence of God based on evil and suffering can be stated as follows:

1. If God exists, by definition God is perfect, all good, all loving, all knowing, and all powerful.
2. Such a God could not do, or permit the occurrence of, any evil or suffering to any other being.
3. However, evil and suffering occur and really exist.
4. Therefore, God, as defined, does not exist.

This does not disprove the existence of a god that is not perfect, one that is not all good, all loving, all knowing, and all powerful, but this is not the kind of personal God that Collins and most Christians deem worthy of worship.

Collins discusses the two kinds of suffering: suffering caused by "what we do to one another,"[21] and suffering that follows "natural disasters, tsunamis, volcanoes, great floods, and famine."[22] Collins says the first kind arises because God gave us free will. However, the existence of "free will" is the subject of major philosophical debate and an unresolved issue in cognitive neuroscience. I am persuaded by the philosophical arguments offered by strict determinists who believe free will is an illusion based on the way our brains function. This view is also consistent with data being collected by brain scientists. I am not going to go into a lengthy discussion about the existence of free will here. Let's say it does exist. What kind of God would give it to human beings? What does this choice say about God's nature? Is God excused from the evil that results?

## FREE WILL AND GOD'S FOREKNOWLEDGE

As C. S. Lewis says, "Nonsense remains nonsense, even when we talk about God."[23] The God described by most religions is all powerful and all knowing but is incapable of logically impossible and inconsistent things. This would mean that God can do anything possible, and knows everything that happens from the beginning of time until the "end times." But God cannot do what is logically impossible or inconsistent. God cannot create a rock so heavy that God cannot move it. There is no such rock. It is not a limit on God's power that God cannot create nothing, that is, the logically impossible rock.

All of that said, wouldn't God have to know ahead of time whether or not you are about to use your free will to cause suffering to another innocent human being? With humans, our thoughts and actions are compartmentalized and occur in sequence over time. Our *knowing* that something will happen, such as that the sun will rise, is not the same as our *causing* it to happen. Our thinking about buying something precedes the action of paying for it. But God does not have separate functions. God's knowing and acting and loving are not separate abilities or sequential mental states. If God is outside of time, as Collins claims, these actions are simultaneous, not sequential. If God is as powerful as most religions assert, God knowing something will happen is the same as God causing it to happen. If humans truly have free will, which is not subject to God's power or knowledge, then God would not know what is about to happen until we acted. God would be unaware of what would happen in God's own universe. God would be a prisoner of time. This is illogical. How is it possible for an all-knowing and all-powerful God and real human free will to exist at the same time?

Another argument against the existence of a perfectly good God is that to give man the "gift" of free will, God would have to have incredibly poor judgment. An all-knowing God would know that some humans would cause innocent humans to suffer. A God that truly loves humans would never give a "gift" that God knows will be so misused. God's love for mankind is surely greater than my love for my children, but I would not give my children loaded automatic rifles to play with. I would have the foresight that sooner or later someone would be hurt or killed. Free will is the most dangerous gift God could give. God could not give this gift without taking responsibility for the inevitable consequence: human-caused suffering. If you accept that free will is a gift from God that allows you to inflict suffering on animals and other humans, didn't God directly or indirectly cause that suffering by giving you free will?

Many believing philosophers have used tortured logic and sophistry to justify how a loving, all-powerful God can coexist with evil and suffering.[24] Many of them claim that human free will was such a good thing for humans that it excuses God from any suffering, no matter how horrendous. Some claim that God has justifiable reasons for permitting evil. These answers do not solve the problem since we still end up with a God that is less than perfectly good or that is limited in power.[25]

Believers argue that without free will, there cannot be virtue in our universe. They also claim that God could not create humans who could be good without allowing evil in their creation. They say that you must suffer grief in order to show sympathy. You must be attacked to display courage. But is this true? Couldn't a truly all-powerful God create a universe in which there was no need to respond to suffering or evil? A world in which creatures are made to be honest, loving, and sharing without the need or capacity to cause suffering? If you lived in that sort of world, you might never need to be courageous or compassionate. Would such a universe be better than what exists?

I would answer clearly, yes. It would be a universe without suffering, and for me, worth foregoing my free will. I would be unable to take credit for my loving, sharing, cooperative actions, but I would be happy to live in a good universe, free of suffering and evil. It is hard to imagine a world in which you have no control over your actions. That's because you want to take actions to protect yourself from pain and evil. If we didn't need protection from either, we would never miss free will. So, a universe that contains creatures with free will is not better than one that does not. The first is a universe with a mixture of good and bad, and the second is a universe that is totally good. If free will was such an important, valuable thing for God, does this mean that the disembodied souls in heaven have free will? If they do really have free will, then it is possible for them to do evil. If they don't have it, then God doesn't regard free will as such an important and valuable thing and is not required to give this gift to humans. God has the power to create humans without free will.

Collins's objection to a universe without free will is that it would be uninteresting. How does he know? There is no reason such a world could not be interesting and fulfilling. Collecting facts and learning things about the universe is interesting and could be done without free will. Experiencing music, poetry, drama, making new friends, and making love would still be interesting even if we enjoyed them without making a conscious choice. Moreover, would a perfectly good God be more concerned with creating a universe with interesting but evil things and people, or creating a perfectly good universe? How much do you treasure the freedom to make mistakes and to cause pain or misery for yourself and others? Do you enjoy the burden of guilt that free will imposes?

If God caused the big bang, created the planets, life-forms, and humans, he also has to cause everything humans do. Their growth, digestion, reproduction, thinking, and feeling have to follow his laws. But suddenly God introduces free will, and God frees humans from the lawful chain of causation? God allows them to cause suffering, for which he is not the direct cause and is powerless to prevent. This also begs the question: are there other things he does not control, and is there a still more powerful god that does actually have power to control everything? Is God not the first perfect uncaused cause (creator) of the universe?

Believing philosophers try to maintain that God can still be called all good if the total good in the universe exceeds the total bad. But, again, this is illogical. All good, perfectly good, means God does not permit, create, or cause *any* evil. Zero. There is no moral balancing act as there is with humans. You permit your child to suffer the pain of an injection for the sake of a greater good. The antibiotic will save the child's life. But you only permit your child to suffer pain because you don't have the power to prevent it and still guarantee a good outcome. If a painless spray method were devised, you would certainly use it.

A God of unlimited power does not have to permit any evil. God has the power to ensure good and eliminate suffering. Are these philosophers saying God's power is unlimited and yet he has to use suffering to obtain good things? This is not the kind of all-powerful God worthy of the love and worship that Christians describe.

Modern philosophers, having been unable to refute the strong argument that the existence of evil and imperfection contradicts the simultaneous existence of a perfectly good God worthy of worship, have tried to weaken it by changing the way the argument is stated. These believing philosophers argue that God has "good reasons" for causing suffering and can still be perfectly good. This philosophical game implies that God develops good reasons to do or permit evil actions. This makes God too much like a human being. God does not have to reason, to think about and justify the consequences of an action before acting. God has to be perfectly good and he acts perfectly, so God's action needs no divine reasoning to justify it. God's action is intrinsically good.

Even if you were willing to allow the possibility that God has to justify action, you have to ask the question, does God have good enough reasons? For humans, goodness is measured by the total effect of our actions. If ten people kill and eat one of their number to survive, we might excuse the suffering this creates for the dead person's loved ones. The survivors had no other alternative. God has no alternative either. God must act consistently with God's perfect nature. God must act without causing suffering if God loves his creatures. It is impossible to conceive of or imagine any remotely plausible reason for any all-powerful, all-loving, all-good God to be compelled to cause suffering.

The believers respond by pointing out that God has greater knowledge than his hapless creatures and can hide his reasons from us. Isn't that deception? Doesn't God risk losing our trust? Do you think there might be divine reasons why incest, torture, and rape are actually good, if we only were able to understand the mind of God and see his hidden reasons? How can we know that God's reasons really justify evil? Or do we just accept this without evidence on faith?

You must concede that horrible, pointless suffering is a frequent occurrence in our world. On NPR I heard a Tutsi woman being interviewed about her experiences during the 1994 genocide in Rwanda.

After seeing her husband and adult son killed, she took refuge in a church. A Hutu soldier found her and cut both Achilles tendons so she could not walk. He then, in spite of the fact she was eight months' pregnant, raped her repeatedly over the next few days. Finally before leaving he stabbed her in the belly, presumably in an attempt to kill her fetus. She was left without food or water but eventually gave birth only to watch, too weak to intervene, while feral dogs ate her baby. What does this say about an all-good, loving God? If the suffering had a point, what would it be? Most of us have watched a loved one, who was a kind and compassionate person causing no harm to anyone, experience suffering or death. Did God want to punish you by causing suffering for your innocent loved one? If God intended to bring your loved one to a state of heavenly bliss, was there a way to do it without suffering? How far do you have to go, and can you ever reach a point, where you can say you have reasonably explained and accepted this undeserved, pointless suffering as justified?

Some Christians don't try to justify suffering or excuse God—they try to make suffering more bearable by claiming that Jesus is there to comfort sufferers and suffers with them. Even if this were true it does not resolve the basic contradiction of evil in God's universe.

Collins and Lewis use the "no pain, no gain" argument. How could we be noble if we could not nobly bear our own suffering or prevent the suffering of others? There can be no good without evil, right? (This is not true. You can experience pleasure without pain. You can be virtuous in a society devoid of vice.) Again, this is an argument for a God with limited power.

But Collins's God is still not off the hook as the cause of suffering for humans, since mainstream Christian doctrine states that God sends human sinners, but not other creatures, to suffer hell. This must cause believers a lot of concern since, if you accept the Bible as true, most of humanity is destined to end up in hell. "There shall be weeping and gnashing of teeth. For many are called but few are chosen" (Matthew 22:13–14). "Then said one unto him, Lord, are there few that be saved? And he said unto them, 'Strive to enter in at the strait gate: for many, I say unto you, will seek to enter in, and shall not be able'" (Luke 13:23–24). Christians need to believe in hell so that evil can be punished and there is justice, even if it has not been obtained in this life. But hell is inconsistent with God's infinite power of forgiveness and mercy. If God and his apostles have power to forgive sin, why isn't *all* sin forgivable?

Collins describes a personal experience as an example of "pure evil," the sexual assault of his daughter.[26] The rapist was never found. But if sentencing were left to Collins, would he sentence the rapist to hell? I would not ask or expect him to forgive or show much mercy, but an eternity of unimaginable pain seems unjust. Yet Christians defend hell by claiming that the evils, the sins of the damned, deserve this punishment.

What sins get you into hell? Well, if you use God's definition of sin as written in the Bible as your guide, something like writing this book (blasphemy), or not doing anything special for the Sabbath, whether the true Sabbath is Friday (Islam), Saturday (Judaism), or Sunday (Christian), will do it. I submit that hell is only needed to restore the sense of justice lacking in an imperfect universe where the evil can prosper and the good suffer. Hell is not logically necessary in a perfect, painless, just universe, and who would claim that God was not powerful and wise enough to create such a perfect universe?

In fact, according to the Bible, God has created a perfect universe. Genesis 1 concludes with verse 31: "And God saw everything that he had made, and, behold, it was very good." There was no evil in the initial creation of the Garden of Eden. Adam and Eve were initially sinless. If we believe the Bible, it is logically and practically possible for God to create a universe that, like the Garden of Eden before the fall, does not contain suffering. But wait—there's more! Christians also believe that the purpose and meaning of life is to live in accordance with the teachings of Jesus and to be rewarded by joining him in the Kingdom of Heaven. Heaven is a state of existence where everything is perfect. Is there any suffering in heaven? Obviously not: "God will wipe away every tear from their eyes. There will be no more death or mourning, wailing or pain" (Revelation 21:4). So, again, God has the power to create a state of existence, a creation, like Eden or heaven, with humans and without suffering. There is no logical necessity that forces God to permit suffering in any universe that God creates.

Another variation on this argument is that God uses suffering to get our attention or to teach us a lesson. Lewis, as quoted by Collins, says God uses suffering as "his megaphone to rouse a deaf world."[27] Was the attack of 9/11 a wake-up call, as some Christians claim? Can't a loving, wise God find a painless way to get our attention, like appearing on every TV, or writing his message on the moon? Can't God find a painless way to tell us, unambiguously and authoritatively, what God wants us to know? Why does God have to be so hard to find and so difficult to understand? Again, as a father, I can teach my children and get their attention without causing suffering and I am not as powerful or as loving as God. What was the divine lesson of my excruciating sciatica? What was God saying to the mothers whose babies were swept out to sea in the Indonesian Tsunami of 2004? Again this explanation of evil limits God's power, and/or God's perfect goodness.

Collins finally gives up any claim of being a reasonable scientist when he says, "we may never fully understand the reasons" for suffering as part of God's plan.[28] What kind of God expects us to live according to a plan that makes no sense to us and is beyond our comprehension? What kind of God would give us a brain that can reason and follow logic then expect us to believe in and worship an irrational, unintelligible, or evil God? Christians can claim the universe has a design, but certainly not an intelligent or intelligible one. Is this the kind of personal God that you are willing to blindly worship?

## NATURAL EVIL

Maybe you are still willing to say that in spite of the man-made suffering in the world, God is still all powerful yet perfectly good. But there is another kind of suffering, the suffering caused by nature: earthquakes, floods, volcanic eruptions, hurricanes, fires, AIDS, cancer, malaria, and so on. Sinners and sinless alike are powerless to prevent these so-called acts of God. How can a perfectly good and loving god cause such horrendous suffering to his beloved humans?

Believers offer a few answers. One is that God is not all powerful. There are some logically possible things that he just cannot do. This is the basic response of the popular book *When Bad Things Happen to Good People* by Rabbi Harold S. Kushner.[29] In my opinion, this destroys any concept of an all-powerful God that created the universe. Why pray to or worship an impotent demigod?

## EVIL IS THE WORK OF THE DEVIL

Another answer is that the devil, not God, causes such suffering. Do we have two supernatural beings with equal power controlling the universe? What evidence in the natural universe supports the devil's existence? If God created everything, then God created Satan. Isn't God then responsible for the suffering Satan might cause? If God is all powerful, couldn't God prevent the suffering the devil causes? The only information bearing on the existence of Satan is, again, the Bible. This assumes the Bible is the true word of God, and we still have yet to show the Bible is more than a man-made book (see chapter 6). T. J. Wray and G. Mobley in their excellent book, *The Birth of Satan*,[30] trace the creation and evolution of this mythological supernatural being. They begin with the rare references to the tempting serpent in Genesis and the adversary that unsuccessfully competes with God in the Jewish Bible. Then they document the evolution of Satan as the all-important personification of evil working in the New Testament.

The Jews' gradual adoption of monotheism created a problem. If Yahweh was the source of everything and was responsible for the benefits, rescues, and victories of the Jewish people, was he not also responsible for the evils, defeats, and disasters they suffered? To quote

Wray and Mobley:

> The existence of an evil deity opposed to the purposes of God eased (but did not finally solve) the tension between the divine power and divine goodness. If there was a Devil, then God was not the author of evil; evil had its own independent source. Certainly this style of dualistic thinking touched on something that is apparently universal and thus applicable to ancient Jews; namely, the binary narrative frame that exists in the human mind, that separates reality into the opposing categories of Us and Them, Friend and Foe, Family and Stranger, and Hero and Villain. To put it simply, the Devil makes for a good story.[31]

As Elaine Pagels points out in her book *The Origin of Satan*,[32] first-century Christians adopted this concept and turned it against the Jews. These Christians characterized themselves as the sons of God opposed and thwarted by the sons of Satan, who rejected and killed their own true Messiah. Nowadays, the devil is considered the tempter who encourages humans to violate the Moral Law and is also the source of cruel catastrophes and uncontrollable personal disasters. But the devil remains a creature of God and God remains responsible for whatever God permits the devil to do. God remains the ultimate source for both the evils of mankind and the evils of nature.

## A BIBLICAL JUSTIFICATION OF SUFFERING

What does the Bible teach us about the problem of suffering? The unknown author of the book of Job attempts to justify evil and the suffering of blameless people. In this story Job is selected as a test case in a disagreement between Yahweh (God) and Satan. (Why a just and all-powerful god would seriously engage in a debate or test with Satan is never explained.) Satan enters into a wager with God, positing that Job, a righteous and God-fearing man, faced with physical and emotional pain, will abandon his faith. God permits Satan to cause Job horrible suffering, even though God attests that Job is a good and blameless man. The suffering includes harming Job's children and wife, who appear to be no more than pawns in this supernatural chess game. Job's patience and faithfulness to God is finally exhausted and he puts the question directly to God: What have I done to deserve such suffering? Why do you afflict a blameless and faithful servant of God?

Here is God's chance to defend himself against this serious charge made by atheists like me and to tear down this barrier to accepting him. The author's pen, guided by the hand of God to write the divine answer, responds: "I am the Lord. How dare you ask for justification of suffering?" God makes the Moral Law but is not compelled to observe it!

The book of Job is great literature with interesting characters and a thought-provoking plot, but in the end, it is another failed attempt to reconcile suffering and evil with a perfectly good and all-powerful deity. What the book of Job does is point out that if you accept Christian dogma, there is no global or general answer to the question of how a good and all-powerful God could justly cause or even permit unjust suffering.

According to Christians, God has a providential plan for everyone. I contend that even if one person has suffered, regardless of the reason, then God cannot be perfectly good. Suffering is not the absence of pleasure. Suffering is a real state of affairs and its mere existence is enough to prove that the personal God of the Christians does not exist.

## MIRACLES VERSUS SCIENCE

Collins feels that the fourth barrier, which causes most scientists to reject the idea of a personal God, is that they cannot reconcile miracles with science. Collins defines a miracle as "an event that appears inexplicable by the laws of nature, and so is held to be supernatural in origin."[33] In defining miracles, Collins defines the supernatural as occurrences beyond our understanding or perception.

As a scientist who deals with perceiving and understanding the real, natural universe, there is no way I can prove the supposed existence of a supernatural realm unless it in some way affects the natural universe. (I discuss the evidence for the supernatural in chapter 8.) Collins's definition only highlights that science has had a very limited amount of time—a little over a thousand years—to explain the many events that have occurred in a complex universe over billions of years. Collins should add to his definition: "appears inexplicable by the *current knowledge of* the laws of nature." It is very possible that with time all alleged miracles will be explained. As science expands, the supernatural contracts. Many events thought to be supernatural at one time are now understood as natural phenomena. Is this the God of the gaps again? It is curious that in prescientific times, reports of miracles were much more frequent.

One of mainstream Christians' core beliefs is in the miracle of Jesus dying on the cross and then rising from the dead. Any doctor or biologist would be taking a huge leap of faith to believe such a claim without extraordinary proof. As it stands, there is no credible, independent evidence that such an event ever occurred. This is where Collins's claim that science and religion are compatible breaks down once again. The biblical accounts are biased, unreliable reports that create a legend about a miracle-worker Jesus. Before science is asked to validate miracles, we need sufficient evidence to establish that the event that you propose to call a "miracle" occurred.

Scientists and people with common sense in general do not have a problem with miracles that they cannot explain; they have a problem with stories about *alleged* miracles. Collins admits he never witnessed or investigated a miracle but he believes stories about miracles. These stories report one-time events of a unique and unusual nature that seem to violate currently known laws of nature. But before we begin to explain a miraculous event, we have to know that it in fact occurred. Christian Bible scholar F. F. Bruce, one of Collins's sources, admits: "So the question of whether the miracle stories of the gospels are true cannot be answered purely in terms of historical research. . . .

The question whether the miracle stories are true must ultimately be answered by a personal response of faith—not merely faith in the events as historical but faith in Christ who performed them."[34]

Once it has been established that the "miraculous" event did in fact occur, then you can begin looking for possible natural explanation for it. C. S. Lewis, the imaginative defender of Christianity, writes:

> Every scientific statement, however complicated it looks, really means something like, "I pointed the telescope to such and such a part of the sky at 2:20 a.m., on January 15th and saw so and so, or I put some of this stuff in a pot and heated it to such and such a temperature and it did so and so." But why anything comes to be these at all and whether there is anything behind the things science observes—something of a different kind, this is not a scientific question.[35]

He proves his point by a self-serving definition. Science, he says, investigates and explains the natural world in which we live, but religion tells us there is a supernatural world that science can never find. So believers in miracles by dogma and definition deny science the possibility of validating them. There can be no scientific evidence, let alone proof of a God, from science. But the existence of a supernatural reality is exactly what is in question. What evidence do we have of a supernatural being? Why miracles, of course. One imagined fact is used to prove another imagined fact. This may be good religion, but it is not compatible with good science.

## WHY ARE MIRACLES SO RARE?

According to Collins, miracles do not occur all the time, but his explanation for God's timing of miracles is nonsensical. He quotes C. S. Lewis: "[Miracles] come on great occasions: they are found at the great ganglions of history—not of political or social history, but of that spiritual history, which cannot be fully known by men."[36] Neither Collins nor Lewis expound on what is meant by "spiritual history," which is not surprising since they are men, not spirits, and cannot possibly know what it is. They seem to be saying that miracles are unpredictable but occur only when God wants to make an important point. Then Collins refers to John Polkinghorne, a believing physicist, who says, "Miracles must convey a deeper understanding than could have been obtained without them."[37]

If miracles are always performed to teach a lesson or to increase our understanding, then the lesson should be clear. But there are many miracles that do not increase our understanding and have no clear, easily determined purpose. Are we supposed to understand miracles or not? Again, in a book arguing the compatibility of God and science, Collins asks us to deal with an inscrutable God beyond our understanding.

In other words, he offers absolutely no explanation for the rarity of miracles at all. As a consequence of miracles being so rare, they are not much use in proving the existence of a spiritual history or God. Why doesn't God put a clear label on these miraculous events to establish beyond a doubt that this was supernatural power at work?

Let us further examine Lewis's claim that God only uses miracles to make an important point in "spiritual history," and that they enhance our understanding of some divine lesson or message. Mark 11:14 and Matthew 21:19 report that Jesus was hungry and spotted a distant fig tree. He came expectantly to eat but found no fruit. A disappointed and angry Jesus curses the fig tree, which withered and died. What momentous juncture of spiritual history necessitated this use of divine power in this circumstance? What deeper understanding of God's purposes or nature or his expectations of humans was realized? Is the message and meaning clear and unmistakable? Clearly, believers and theologians have used their creative imaginations to torture out multiple interpretations and to create all kinds of divine messages in dissecting the texts describing this miracle. This is only confirmation of my contention that the lesson, if any, in the text is obscure and ambiguous.

Despite Collins's claim that miracles "should have some purpose rather than representing the supernatural acts of a capricious magician, simply designed to amaze,"[38] this fig tree story represents just such an act. There is no lesson in the text attributed to Jesus. Given the imagination and creativity characteristic of believing Christians, I'm sure some message, in fact, many different messages, could be metaphorically read into the text. But on the face of it the literal text does not clearly support any of the numerous subjective interpretations. What was the spiritual historical significance of changing water into wine and what lesson does it teach us? Many additional examples of apparently pointless miracles could be cited.

If Jesus performed miracles to establish that he was a divine messenger on a divine mission, then these supernatural demonstrations were not convincing to the Jews and Romans of his times. They witnessed his miracles and still treated him as a troublesome man, or a blasphemer, and a false prophet. These were people with limited scientific knowledge who readily accepted the supernatural. If all the supposed eyewitnesses of Jesus' alleged miracles, including the resurrection, were not convinced of Jesus' divinity, why should anyone, scientist or nonscientist, accept miracles on the basis of thirdhand biblical stories today?

The end of the age of miracles and the rarity of miracles nowadays remain to be explained and justified. It is amazing that Collins does not recognize that belief in miracles could frustrate and impede the basis of any scientific experiment, even his own. Imagine Collins conducting an inconclusive experiment, or one that produces results inconsistent with accepted science. What would he say to a believer who wanted to label the results a miracle? Such an explanation is no more acceptable than attributing the result to magic. Even if you are willing to entertain the possibility of miracles, you would have to admit that on the basis of experience they are very, very unlikely. It would require more than eyewitness reports and take a lot of investigation to rule out all possible natural explanations.

# NOTES

1. F. Collins, *The Language of God* (New York: Simon & Schuster, 2006), p. 34.
2. Ibid., p. 35.
3. Ibid., p. 38.
4. Ibid., p. 37.
5. C. S. Lewis, *Surprised by Joy* (New York: Harcourt Brace, 1955).
6. Collins, *Language of God*, p. 36.
7. Ibid.
8. Ibid.
9. Ibid., p. 38.
10. Pew Forum on Religion and Public Life, http://www.pewforum.org/news/display.php?NewsID=16260.
11. D. E. Comings, *Did Man Create God? Is Your Spiritual Brain at Peace with Your Thinking Brain?* (Duarte, CA: Hope Press, 2008).
12. D. C. Hamer, *The God Gene* (New York: Anchor Books, 2005).
13. D. E. Comings et al., "The DRD4 Gene and Spiritual Transcendence Scale of the Character Temperament Index," *Psychiatric Genetics* 10 (2008): 185–89.
14. M. Persinger, *Neurobiological Bases of God Beliefs* (New York: Praeger Publishers, 1987).
15. V. S. Ramachandran and S. Blakeslee, *Phantoms in the Brain* (New York: William Morrow, 1998).
16. A. Newberg, E. D'Aquili, and V. Rause, *Why God Won't Go Away: Brain Science and the Biology of Belief* (New York: Ballantine Books, 2001).
17. Collins, *Language of God*, p. 40.
18. E. Gunn, "Death by Prayer," *Freethought Today*, September 2008, pp. 6–7.
19. L. Gayler, *Woe to the Women: The Bible Tells Me So* (Madison, WI: Freedom from Religion Foundation, 1981); P. Christ, *Womenspirit Rising: A Feminist Reader in Religion* (New York: Harper One, 1992); E. Stanton, *Women's Bible* (Amherst, NY: Prometheus Books, 2005).
20. http://drbarrywhitney.com.
21. Collins, *Language of God*, p. 43.
22. Ibid., p. 44.
23. Quoted in ibid., p. 43.

24. A. Plantinga, *God, Freedom and Evil* (New York: William B. Eerdmans, 1974); R. Swinbume, *Providence and Problem of Evil* (New York: Oxford University Press, 1998); M. McCord Adams and R. M. Adams, eds., *The Problem of Evil* (New York: Oxford University Press, 1990); D. Howard-Snyder, *The Evidential Argument from Evil* (Bloomington: Indiana University Press, 1996).

25. For atheist rebuttal see M. Martin and R. Monnie, eds., *The Impossibility of God* (Amherst, NY: Prometheus Books, 2003).

26. Collins, *Language of God*, p. 44.

27. Ibid., p. 46.

28. Ibid.

29. H. S. Kushner, *When Bad Things Happen to Good People* (New York: Avon Books, 1983).

30. T. J. Wray and G. Mobley, *The Birth of Satan* (New York: Palgrave Macmillan, 2005).

31. Ibid., p. 167.

32. E. Pagels, *The Origin of Satan* (New York: Random House, 1995).

33. Collins, *Language of God*, p. 48.

34. F. F. Bruce, *The New Testament Documents: Are They Reliable?* (Grand Rapids, MI: William B. Eerdmans, 1981), p. 67.

35. C. S. Lewis, *Mere Christianity* (New York: HarperCollins, 2001), p. 22.

36. Collins, *Language of God*, p. 53.

37. Ibid.

38. Ibid.

*Chapter 4*

# WHAT'S WRONG WITH THE MORAL ARGUMENT?

> *Man is constantly inflicted with a defect . . . the Moral Sense. It is the secret of his degradation. It is the quality which enables him to do wrong. Without it, man could do no wrong. He would rise at once to the level of the higher animals.*
>
> —Mark Twain

## THE MORAL LAW: DOES IT EXIST?

Collins and his mentor C. S. Lewis argue that in matters of human behavior, everyone uses a standard of some sort to classify behavior as right or wrong, acceptable or unacceptable, good or evil. Collins believes that the existence of a universal "Moral Law" offers the most compelling evidence that a divine lawmaker exists. Briefly stated, he argues (1) the Moral Law exists in all human beings, (2) science cannot explain the Moral Law by the natural processes of either a genetically inherited behavior or a culturally acquired ability, and therefore (3) the only explanation left to fill the explanatory gap is a supernatural one, namely, the God of the gaps.

Typically, Collins and Lewis make their point using analogies and metaphors. But in so doing they carelessly cite behaviors that are not essentially moral or intrinsically good or evil. As discussed in chapter 3, analogies can be used between two things that are very similar with respect to the characteristics compared. Analogies and metaphors never provide proof positive—only reasons to consider the likelihood of the comparison being true. Lewis offers football rules that establish whether a foul has been committed as an invalid analogy. Football rules are not a part of human nature, nor are they divine commandments. They are social conventions. They exemplify one of many arbitrary, man-made standards for behavior that are designed for a given set of circumstances. Other examples of man-made rules include etiquette, traffic laws, sports, games, and so on.

Many behavioral rules, such as initiation, marriage, burial rites, the use of the metric system of measurement, and so on, are the result of cultural consensus. In these examples, we are talking about what is deemed "right or wrong" or "acceptable or unacceptable" by some culture in some defined situation, not about a Moral Law. But if you are going to argue that there is a universal Moral Law, it cannot be a man-made rule. It has to be a universally accepted "Law of Good and Evil." But is there such a thing? Collins and Lewis want to relabel this law of universal good and evil the "Law of Nature" because they believe it is an integral part of human nature. So, Lewis says, "this law was called the Law of Nature because people thought that everyone knew it by nature and did not need to be taught it."[1] But what is the "it" that everybody "knows"?

Let us look closely at "it," the Moral Law, in terms of describing something that exists in nature. It is a fact that human beings feel that certain behavioral standards apply in group situations. The capacity to set these behavioral standards (based on how your actions affect other human beings) exists in our brains at birth. Yes, human beings have a natural imperative to make rules regarding social behavior. These rules are based on our emotional reactions to behavior and on what we and members of our group find acceptable and learn to expect from each other. But is there a single set of universally good and evil behaviors self-evident to all human beings that can be covered by the term *Moral Law* as used by Collins in his discussion of evidence? No, there is no universal Moral Law in this sense. Humans acknowledge that there are rules that define good and evil. But not all human beings accept the same set of rules.

There is a distinction between a natural or universal need to judge behavior and the specific criteria we use when we exercise that judgment. We can agree with Collins that human beings have a sense of what they are supposed to do in a given situation. It is an instinct I prefer to call our *moral sense* rather than our innate knowledge of and agreement with a universal Moral Law.

## IS THE MORAL LAW A PRODUCT OF CULTURE?

Having established that human nature includes a moral sense, Collins asks whether it is "an intrinsic quality of being human or just a consequence of cultural traditions?" Collins and Lewis recognize that a major objection to their belief in an inherited set of universal standards is clear evidence that different civilizations and different ages have viewed good and evil very differently. Their response to this objection is that this claim is "a lie."[2] Any differences over time and between cultures, they claim, "have never amounted to anything" significant. A lie is the conscious act of one or more humans to state as true what they know to be false. Collins and Lewis point out that in all religious or ethical codes throughout history there is one "monotonous denunciation of oppression, murder, treachery, and falsehood; the same injunctions of kindness to the aged, the young, weak."[3] But the definition of these terms—oppression, murder, treachery—depends on your point of view. Do the Israelis oppress the native Palestinians or do they just protect their own population? Are abortions murders? Were Hiroshima and Nagasaki war crimes or self-defense? Was the Trojan horse treachery or clever strategy? Collins and Lewis insist on looking at a few common generalizations and innate natural instincts and ignoring the many, many variations and exceptions when these generalizations are applied to specific actions and behaviors. While Collins and Lewis are entitled to their opinions and interpretations, they are not entitled to make up their own facts.

The fact is, there are many and varied versions of good and evil in any cross-cultural or time-related study. To mention one instance of moral evolution: for many centuries human slavery was not regarded as evil. Authors of both the Jewish Bible (which Christians call the Old Testament) and the New Testament accepted it. During the Civil War not a single member of the Catholic hierarchy condemned slavery. This is a significant moral issue. Behavior that is punishable by the death penalty has changed over time and within different cultures. This is a significant issue. Indeed, the ongoing effort to define universal standards for good and evil is a major, unresolved project of the branch of philosophy known as ethics.

If moral standards are such an essential part of human nature, wouldn't these standards be self-evident and universally accepted? Wouldn't the study of ethics be redundant? Wouldn't cultures over time gravitate to the same standards? Ethicists cannot agree on what rules to use to judge the goodness or evil in an action. Should it be judged by divine law, by some set of reasonable rules, by the actor's intention, or only by the consequences of the action?

Collins concedes that there may be variations in the content of the Moral Law, for example, "In some unusual circumstances the law [of good and evil] takes on surprising trappings—consider witch burning in seventeenth-century America."[4] Consider witch burning, indeed! It was not confined to seventeenth-century America. The torture and killing of witches is an ancient practice that preceded the biblical command, "Thou shall not suffer a witch to live" (Exodus 22:18). The papally commissioned report on rules for witch trials, "Malleus Maleficarum," was published in 1486, and updated revisions were published continually until 1669.[5] Collins should not so lightly dismiss the failure of the Law of Good and Evil to stop the killing of tens of thousands of witches over several centuries. Over eighty thousand mostly elderly women were killed in Western Europe between 1480 and 1690.[6] The last reported execution of a witch in Europe was in Ukraine, in January 1997, and a witch was executed in Kaskaskin, Illinois, in 1870.[7] As I write this, a woman in Saudi Arabia is being tried on charges of witchcraft and she faces the death penalty. In Nigeria, the Christian Liberty Gospel Church, led by witch-hunter Helen Ukpabio, used arson, beatings, and execution against children and adults for alleged witchcraft in 2009.

Collins's response is that Moral Law, instilled in us all by God, requires that we kill witches if we really believe they do horrible things and are servants of the devil. I'm sure this argument would be accepted by the perpetrators of the 9/11 terrorist attacks, who sincerely believed we Westerners are servants of the devil and should be killed for the horrible things we do. I do not believe that Collins would consider the murder of a compassionate physician who terminate pregnancies at the request of distressed women as justified on the basis that the murderers considered the physician "the personification of evil on earth, an apostle of the devil himself."[8] Besides witchcraft, what other horrible things would justify the death penalty in the minds of believers like Collins? Blasphemy? Adultery? Defiling the Sabbath? Homosexual behavior?

The fact is that Collins acknowledges the Moral Law is not clear, uniform, or universal. Some people accept killing people for certain reasons, and some do not. Some accept killing in self-defense as moral, and some do not. Some claim that killing in a just war is acceptable, and some say there is no such thing as a just war. Some defend the death penalty for those who commit horrible crimes and treasons, and some do not. All of these people with differing opinions share the same human nature but seem to lack a universal supernatural guidance to unfailingly tell them what is good and what is evil when it comes to killing other humans. Clearly, culture-nurture affects the Moral Law to some extent. Think about how you learned what behavior was good or bad, that is, acceptable to your parents and their community. Wasn't this also culture dependent? How different would your moral standards be if you were born to a family of Islamic nomads in the Saudi Arabian desert?

I am baffled why Collins was "stunned" by C. S. Lewis's logic when he wrote, "If there was a controlling power outside the universe [*something yet to be proved*] it could not show itself to us as one of the facts of the universe . . . [*Why not?*] The only way in which we could expect it to show itself would be inside ourselves as an influence or command trying to get us to behave in a certain way. And that is just what we do find inside ourselves. Surely this ought to arouse our suspicions."[9] This is not a statement of stunning logic, but hypothetical speculation.

I would suggest that Charles Darwin gives a more logical explanation of this voice of conscience within us. In his book *The Descent of Man*, he wrote, "It is worthy of remark that a belief constantly inculcated during the early years of life, whilst the brain is impressible, appears to acquire almost the nature of an instinct and the very essence of an instinct is that it is followed independently of reason."[10] There is a 7 in 10 chance that you have the same religious affiliation as your parents. The voice of conscience is developed in you by these early cultural influences.

## IS THE MORAL LAW NATURALLY INHERITED?

Doesn't Collins have a point that cultures share many of the same moral standards and tend to reward or punish for the same behaviors? Yes, he does. There are common restrictions on killing and having sex with family members and rules for making and honoring promises. Is the only explanation, the only cause of such common moral codes, an act of a supernatural, law-giving God? No, it is not. Common features can be explained by common genes. We and other social animals share common instincts that inhibit killing members of our group. Having sex with close relatives is called inbreeding and can easily lead to extinction, as can be seen in the case of the cheetah. Male lions and elephants are excluded from their family group when they reach sexual maturity, which reduces inbreeding. Studies of social animals are full of behavioral rules and behaviors that evolve and are passed on by what seem to be natural, not supernatural means. The rules or behavioral patterns that are passed on are those that increase the chance of reproduction and survival of offspring.

So, perhaps the common elements of Moral Law are the result of both the way our animal brain is genetically formed and operates, and society's need for rewards and punishments to ensure that it survives and prospers. While humans have developed more elaborate moral rules and behaviors than our fellow animals, we are still part of the natural world and our behavior needs no supernatural explanation. At the very least it seems that genetics—nature—also has a clear effect on the Moral Law.

Collins is an evolutionist (like almost all scientists), so he is concerned that genetics, sociobiology, and evolutionary psychology can offer natural explanations for moral laws. Since such explanations undermine his faith that the moral sense is God's handiwork, Collins focuses on the hardest case for natural science to explain, namely, altruism, or unrewarded generosity.

Altruism, according to Collins, is a behavior that would contribute to your early elimination in the fight for survival. It would not be favored by evolution. In Collins's opinion evolutionary science cannot explain why such a behavior persists in humans. Since there is no natural explanation, Collins seems to think it must have been incorporated into human nature by a supernatural being, namely, the God of the gaps.

## WHAT ACCOUNTS FOR ALTRUISM?

Collins defines altruism as "the truly selfless giving of oneself to others with absolutely no secondary motives."[11] There is a real question whether such perfect altruism exists. I first considered this problem during World War II. I had heard about soldiers who, seeing a hand grenade fall in the midst of their buddies, unhesitatingly jumped on it, absorbing the blast and saving their comrades. This impressed me as pure altruism. However, on further thought, the soldier may well have been acting reflexively, like taking your hand from a hot object, without any conscious decision. Can you always determine the conscious primary and secondary reasons for your behavior? What about unconscious motivations for inexplicable behavior? Even if we can credit the soldier with the primary conscious motive of saving his comrades' lives, we cannot logically rule out secondary or unconscious motives. The bond between soldiers sharing life-and-death situations is extremely strong. The thought of surviving and dealing with the loss of life and serious wounding of his band of brothers may have been more painful than the risk of death itself. This survivor guilt is irrational but well documented. A secondary motive could have been his desire for approval from his peers or for the recognition that heroes and patriots receive. There is no way we can, with certainty, logically infer that the act, however praiseworthy, was pure, conscious altruism without any conscious secondary or unconscious motives.

Similarly, Collins points to Mother Teresa and Oskar Schindler as exceptionally altruistic people. And without question, they deserve high praise for their actions. However, when we offer others benefits that work to our own benefit, this is called *reciprocal altruism*, not pure altruism.

There were limits to Mother Teresa's and Schindler's giving, and both expected some benefit in return. Mother Teresa was motivated by saving her own soul as well as those of Calcutta's poor. She has been criticized for failing to provide modern medical care despite raising more than $50 million.[12] In a recently published book of her personal letters, Mother Teresa confessed to constant doubts about God's existence and a secret unhappy life that was "dry," "empty," "lonely," "torturous," and "devoid of all feeling." All the while, she continued to deceive the world with her verbal expressions of unquestioning faith.[13]

Oskar Schindler was an alcoholic and adulterer who was a member of the Nazi Party, making ammunition for the German war machine. His real motivations for saving so many Jews are unknown and were probably both laudable and to some degree practical. Mother Teresa and Schindler do not illustrate a universally infused predilection for pure altruism, but the all-too-natural mix of inherited behavioral traits, including those that selflessly serve others and those that are self-serving. These mixed motives find different expressions in different circumstances. Do you think the rarity of pure altruism is evidence that it is a universal characteristic that all humans possess? If it exists, is it a possibility for only a few special human beings? Does it seem logical to claim that a rarely, if ever, expressed facet of human nature proves that God created all humans with the potential for altruism as part of his universal Law of Good and Evil?

Most philosophers who study good and evil—ethicists—would agree we have a moral duty to do good and avoid evil. But that duty would not require us to be supergood, altruistic, or, as Collins suggests, put our lives at risk to save a scorpion. There is a religion in India called Jainism that has such a rule, and its practitioners may well risk their lives to save an animal or an insect. But most of us would not be so inclined. Being good does not require completely ignoring our own self-interests for the sake of others.

Are cooperative behavior and altruism uniquely human as Collins suggests? Clearly, social animals demonstrate cooperative behavior, such as hunting and group defense of territory used by lions, wolves, and chimpanzees. What about the tougher problem of altruism? Chimpanzees demonstrate unselfish behavior to help other chimpanzees or humans with no obvious reward or benefit to themselves.[14] Their behavior is the same as that of eighteen-month-old human infants. Both will bring you an object you have dropped and cannot reach without needing to receive a personal reward. In humans, this innate, genetically inherited tendency toward altruistic behavior is strengthened by culture and reinforced by learning over a lifetime. Can we attribute animal cooperation—and yes, even animal altruism—to native genetics and environmental influences, or do we need to resort to the supernatural action of God? Are animals and humans so different behaviorally that we need a supernatural explanation for the latter but not the former? Is the guilt you feel when you fail to behave as you think you should different from the guilt your dog feels when he violates your rules? Is the sense of "should" that higher nonhuman animals feel also instilled by God?

Collins accepts the fact that genes influence behavior. In his appendix, he discusses twin studies citing evidence for inheritance of behavioral tendencies and also discusses the isolation of genes associated with behavioral traits such as novelty seeking and violent behavior. Is there a gene for altruism? Geneticists who study behavior today virtually all agree that there is no single gene that directly determines whether or not you will have altruistic tendencies or any other complex behavior.[15] In almost all circumstances there are many interacting genes that contribute to your behavioral limits. So, are there genes that contribute to altruistic behavior? Richard Ebstein at Hebrew University measured altruism using a test called the Selflessness Scale and a gene that increased sensitivity to dopamine, a brain chemical that makes us feel good.[16] Three out of four individuals with high scores for altruism had the gene. Does this prove altruism is genetic? Of course not, but it does support the fact, acknowledged by Collins, that genes and genetic mechanisms play a significant role in shaping our behaviors and personalities.

Collins accepts the results of a study, cited in his appendix, of the variations in low-activity and high-activity monoamine oxidase A (MAOA) due to a specific gene. MAOA breaks down chemicals that act as messengers in the nervous system, such as serotonin, dopamine, and norepinephrine. Individuals with low-activity MAOA were found to be more aggressive, but the gene alone only accounted for a small percentage of the difference in aggression between individuals. However, a twenty-six-year-old study of males in New Zealand with the low-activity "bad" gene revealed that 85 percent of the males with the gene for low activity who were also subjected to child abuse demonstrated some form of antisocial behavior. This study is only one of many that illustrate that behavior is potentially explainable in terms of natural phenomena, even if we can't predict the behavior of individuals based on genes alone. Ebstein's research cited earlier reported that changes in sensitivity to dopamine, the feel-good neurochemical, associated with high scores on the Selflessness Scale were due to changes in receptor genes.[17] His group recently reported that individuals who had long versions of a gene (vasopressin 1a receptor) scored high on the Value Expressive Behavior Scale, a measure of altruism. When playing the Dictator Game, where the player is given a sum of money and told he can decide how much he wants to give to another player, the individuals with two copies of long versions of the gene gave significantly more than those with two copies of the short version.[18]

Beyond the genetic influence on our brain functions and structure, the natural basis of "altruism" can be studied by directly stimulating or disabling parts of the brain. Ernst Fehr also uses games to study issues like decision making and fairness. There are a number of these games, such as Prisoner's Dilemma, Dictator, and so on, that are widely used in psychological research. For a short, readable summary of game theory, see *The Survival Game: How Game Theory Explains the Biology of Cooperation and Competition*.[19] These experimental games find little or no evidence for altruism, which is rarely displayed. The Ultimatum Game is one that involves two players, one of whom is given a large amount of money, say $100, and is required to offer part of it to the other player. If the other player refuses the offer, they both get nothing. Normally, if the offer is too small, it will be refused as unfair even though rationally any money is better than none at all. When the investigators disabled the right side of an area of the brain called the dorsolateral prefrontal cortex (DLPC) by temporarily sending an electric current through it, the usual tendency to punish unfair or selfish partners was inhibited. The individuals with disabled DLPC took any offer, no matter how unfair.[20]

People with defective brains at birth, or who suffer brain injury, also can lose the ability to know when social rules are broken.[21] So, our sense of fairness and our ability to punish cheaters—violators of the Moral Law—is explainable in terms of brain chemistry and function.

## HOW DOES SCIENCE EXPLAIN MORAL BEHAVIOR?

Collins, after dismissing "nurture" or "culture" to explain the Moral Law, also dismisses "nature," the argument that the Moral Law is "simply a consequence of evolutionary pressures."[22] He accepts the scientific theory of evolution but argues that there is no survival value in nature-based explanations of why humans intermittently follow the Moral Law, especially in the case of altruism. He does acknowledge, "There is an inescapable component of heritability to many human behavioral traits."[23] But apparently he does not consider our moral sense as one of them.

It is surprising that Collins fails to discuss evidence from behavioral genetics or evolutionary psychology. The Human Genome Web site, under the title "Behavioral Genetics," states: "Behavior has an evolutionary history that persists across related species. Chimpanzees are our closest relatives, separated from us by a mere 2 percent difference in DNA sequence. They and we share behaviors that are characteristic of highly social primates, including nurturing, cooperation, altruism, and even some facial expressions. Genes are evolutionary glue, binding all of life in a single history that dates back some 3.5 billion years. Conserved behaviors are part of that history, which is written in the language of nature's universal information molecule—DNA."[24]

Collins does refer to sociobiologists, but claims that their explanations "in terms of indirect reproductive benefit . . . run into trouble."[25] There are a large number of scientists, perhaps the majority, who believe that the field of sociobiology has been generally successful in demonstrating the genetic basis of human nature.[26] Collins also ignores that there is ongoing research along these lines and that evidence of the evolutionary benefits of cooperation and altruism continues to accumulate.

Collins cites infanticide as evidence against altruism being a "positive factor in mate selection."[27] Infanticide—the practice of killing young in the group, observed in monkeys, lions, and rats, among other species—is practiced by males who have successfully replaced the dominant male in their group. However, there are evolutionary benefits of such infanticide. First, by eliminating the genes of the displaced alpha male, only the strongest males pass on their genes. In addition, since nursing females are not receptive to sexual advances of males, the new alpha male converts sexually nonreceptive mothers into sexually receptive mates. This behavior is discussed in a somewhat technical book, *Infanticides by Males*.[28] Infanticide is limited to selective species and is not practiced by humans, so it is irrelevant as an explanation or argument about altruism. A 2007 article, "Altruists Attract," published in the scientific journal *Evolutionary Psychology*, makes a case based on experiments using human subjects that altruism increases mating success.[29]

While Collins argues that evolution only takes place at the level of the individual, some evolutionary psychologists, behavioral geneticists, anthropologists, and neurocognitive scientists would disagree.[30] They would also explain the Law of Good and Evil in terms of evolutionary, genetic, and cultural events working at the level of the group. Altruistic behavior and moral sensibilities could be selected both at the level of individuals due to genes that increase reproductive success and by environmental events that select between groups of differing altruism. As an argument against altruistic or cooperative behavior having an advantage at the group level, Collins asserts, "It would seem to require the opposite response, namely hostility to individuals outside the group."[31] This is precisely what is observed in nature among social animals. Strangers are attacked and groups defend their territory against other groups, whether it's a troop of baboons, a pride of lions, a Washington, DC, street gang, or the Palestinians in Gaza.

The unconscious selection of reproductively successful individuals can occur at the level of the individual or group. Research on this subject is still in its early stages. Nevertheless, it is clear that genes control the formation and structure of our brains. Genes control the quantity, function, and interaction of the metabolic and electrochemical processes that direct brain activity and consciousness. In this sense, certain behavior patterns become part of human nature.

Almost all social animals are instinctively (genetically programmed) against killing blood relatives or members of the immediate social group. This applies to a pride of lions, a troop of monkeys, a pack of wolves, or humans. Cooperation has obvious survival value, including the ability to share living space and food. It leads to genetically programmed, common "moral" behavior that has survival value for both the individual and the group. Rules for "moral" sexual reproduction have survival value by preventing killing that would occur in a rule-free, sexually based competition.

Evolutionary psychologists have collected evidence to show that prohibition of incest has a biological basis.[32] There is, therefore, a plausible genetic explanation for the evolution of cooperation and cultural rules that result in morality among many social animals, including man. This is reinforced by a genetic tendency in social animals to punish rule breakers.[33]

There are several books that offer reasonable natural explanations of the tendency of human groups to establish rules, to reward cooperative moral behavior toward one another, and to punish cheaters.[34] These authors agree that this moral sense is ultimately a natural characteristic and not something so "unusual" that they are desperately looking to explain by inserting the God of the gaps. Scientists today are only debating the details. How much of altruistic or moral behavior is due to nature (genes), and how much is due to nurture (cultural learning)? They debate whether evolutionary selection is at the group or individual level or both. They debate the role of kin selection, reciprocal altruism, cheating detection, punishment, reputation, and so on, in contributing to the moral sense, but they do not debate whether the moral sense is a natural or supernatural phenomena.

## IS THE MORAL LAW AS NATURAL AS SPEECH?

Collins enumerates the special qualities of human beings as "the awareness of right and wrong, the development of language, the awareness of self, and the ability to imagine the future."[35] He does not expand on any of these elements of human nature. But before we accept *The Language of God*, let us consider the language of men.

Perhaps we should apply the analogy of the Law of Good and Evil to language. The ability to communicate by speech is a genetically inherited brain module. It is not a culturally acquired trait. All normal humans have the ability to speak. But does the fact that all humans have a speech module in the brain mean all humans speak the same language? Specific language depends on the environment in which the speaker is raised. It depends on the language spoken in the home, in school, in the community. It is culture dependent. There is a basic structure or grammar common to all languages in every culture. Let us call this the Law of Language. Most scientists agree that the accumulation of experimental and observational evidence demonstrates that the ability to speak a language is a genetically inherited trait in humans. The capacity for language is an essential part of human nature and is developed in, not taught to, human infants.

There is a genetic universal pattern of grammar and syntax, but the meaning of words and the way they are vocalized and arranged are learned. There is a natural scientific explanation for this language instinct, and it does not require a divine origin.[36] Why would we propose that the ability to form moral rules, the Law of Good and Evil, cannot be scientifically explained by natural processes, while the ability to speak, the Law of Language, can?

Again, it's hard to see that this moral sense is entirely different from other natural behaviors and requires a special supernatural cause. The fact that human communication is so well developed compared to animal communication does not mean it is something supernatural, nor does the fact that humans develop complex rules for behavior in groups mean that our rules are supernatural and animal rules are not.

If you can address and refute an opponent's strongest argument, you are very likely to refute any arguments of lesser strength. As for Collins's argument that altruism can only come from God, it seems hard to prove that pure, strong altruism even exists. Such altruism as does exist can be explained by social animals' natural evolutionary developments that contribute to reproductive success today or have contributed to reproductive success in the past.

## IS THE MORAL LAW AN UNINTENDED CONSEQUENCE OF EVOLUTION?

Even if Collins is not persuaded by the considerable evidence and arguments that moral laws have a survival advantage, there is another natural explanation. Specifically, moral behaviors could result from a combination of other mental activities that served an evolutionary purpose in prehistory but are obsolete in modern society.[37] Evolutionary biologist Stephen Jay Gould calls these "spandrels," an architectural term used to describe triangular, useless spaces that unintentionally result when two arches are intentionally put side by side. In the course of evolving a big, complex human brain, our tendency to assign moral values or to practice altruism may have served a different purpose in the past that did have survival value. Over time the function was no longer needed for its original purpose. It may have become an evolutionary functionless leftover, like the human appendix or whales' hind legs, or maybe nature has found a new use for it. This is a possible and entirely natural explanation. Whether or not evidence supports it is the subject of future research.

## OTHER PROBLEMS

As an evolutionist, Collins also fails to specify when this divinely inspired addition to human nature, the moral sense, was first inserted. Did *Australopithecus* appeal to his moral intuitions before stealing food from his sleeping neighbor? What about *Homo erectus* or *Homo habilis*? Did they have moral codes of behavior that were inherited and not taught? If apes have rudimentary moral rules for behavior, then how can we claim we are the result of special creation? There is archaeological evidence indicating that the moral sense appeared some 6 million years ago in a common ancestor and was not suddenly, divinely added into first-century humans.

We also need to consider what Collins and Lewis mean by "law." Again, this is an invalid metaphor. In one sense, they use "law" to refer to explanatory and predictive relationships that describe how the universe operates. These "laws" are self-executing. They cite the laws of gravity and special relativity as examples. These laws apply to all of us all of the time and are not subject to repeal by humans. These laws are almost all about the relationships between nonliving elements of the universe.

Laws of this class are more difficult to formulate for living organisms because of the complexity and the many possible combinations of inherited abilities and environmental interactions. Is the Moral Law a law, in this sense, at all? The Moral Law is violated constantly, so much so that the violations could also be described as intrinsic to human nature. Even the basic drive of living organisms to continue living is not universally observed by complex organisms like humans who commit suicide in significant numbers. What Collins and Lewis characterize as a law of human nature is more accurately classified as a *description* of human nature and has no predictive or explanatory value. In fact, Collins prejudices the argument when he names the phenomena the "Moral Law" because he assumes, without independent evidence, that there is a lawmaker, which is the very claim that he is trying to establish.

A final possibility is that free will, which is necessary for a meaningful Law of Good and Evil, might be an illusion. But that is the subject of another book. Let us suppose Collins is right, that God did create humans with a need to establish standards by which to classify human behavior as good or evil. Do you think God would be satisfied with standards any group of humans freely adopted? This doesn't sound reasonable and makes Moral Law relative, something Collins wants to avoid. Clearly, a personal God would want humans to follow the correct standard, which, of course, is God's standard as expressed in his revelation. Did God give us a revelation? Many works are alleged to be divine revelations. But which revelation, if any, is genuine? For Collins this would be the revelation of Jesus and his biblical teachings.

What kind of world would we expect if Christian standards were the only correct standards for moral behavior? We would predict that Christians, on average, would more frequently follow the right standard and they would, on average, practice more good behavior and less evil behavior than people who followed standards of other religions or no religion at all.

However, analysis of actual behavior does not show Christians to be demonstrably better on average. Dawkins, Hitchens, and Harris cite statistics to show that, in fact, if a difference exists at all, they are *less* good than atheists and agnostics. The time in history when Christianity had the exclusive power to set and enforce standards of behavior—the Middle or Dark Ages—was not exceptionally good, and, in fact, a fair amount of evil occurred. The nation with the most believing Christians, the United States, has a higher rate of incarceration, murders, child abuse, rape, and so on, than the more atheistic nations of Europe. The charges of these critics of religion may or may not be exaggerated, but the point is, even allowing for free will, Christians throughout history have behaved exactly the same as any other humans, irrespective of their religious beliefs. Do you really know how good or evil a person is if I tell you he is a minister or an atheistic astrophysicist?

The argument that Collins and many other believing scientists and theologians use to support the concept of the Moral Law is not rationally acceptable. Logicians call it "the argument from ignorance." It involves first establishing an event or phenomena in nature, call it X (in this case, X represents the moral sensibility that humans demonstrate), then critically rejecting all possible explanations of the event or phenomena. Call these explanations Z1, Z2, and so on (in this case genetics, culture, chance). Explanation Y is then proposed as the real explanation for the event or phenomena.

As an example of how this argument is used, let's consider the case of the husband who arrives home much later than expected. What explains this event? Obviously there is more than one possible explanation. You could investigate and dismiss working late or sudden illness as explanations. Likewise you could dismiss alien abduction as a realistic explanation. Having eliminated these explanations, however, does not justify your accepting the explanation that he is having an affair. It is also possible that he could have had car trouble, or that he stopped to assist a friend in trouble. Dismissing some explanations as untrue or unlikely does not make one of the remaining possible explanations true by default. The event remains unexplained, in the absence of direct supporting evidence of the explanation selected. We could express an opinion as to what the explanation might be, but it would be only an opinion, not real knowledge.

In this case Collins asks for an explanation of the Moral Law. Collins rejects genetic and cultural explanations as not explaining the "Moral Law" to his satisfaction. Then the argument for ignorance concludes X, the Moral Law, must be due to his preferred alternative explanation Y, God did it. But this conclusion is not valid. Even if you disagreed with my evidence that genes and experience explain the Moral Law, all you could validly conclude is that it remains unexplained. No matter how many explanations of the Moral Law you have shown to be false, you have not proved the truth of any of the remaining possible explanations. There are additional possible explanations from genetics or the environment that could be proposed for the Moral Law that could be true. If instead of God, Collins proposed that the Moral Law was due to an undetectable altruistic virus implanted in human brains by extraterrestrials, would we, by rejecting the natural explanations proposed, have to accept this explanation as valid? This is why I find Collins's proposition that the Moral Law is a signpost pointing to divine intervention irrational and not intellectually satisfying.

The existence of moral sense cannot be used in a logical argument to prove the existence of the supernatural. Deductive logic is used to reason from true premises to necessarily true conclusions. Collins's argument is not a logically deductive argument at all. It is an adductive argument: an inference to the best explanation. However, to reach a valid conclusion using the adductive form, you have to show that there is one and only one explanation for the Moral Law, in which case the conclusion is valid. But obviously this is not true of the Moral Law, so this form of argument becomes a process of selecting the best possible and most likely explanation. The conclusion then is probably, but not necessarily or absolutely, true.

How do you decide what makes an explanation acceptable and scientifically, intellectually satisfying? Scientists use criteria such as: How well do the elements of the explanation relate to the phenomena? Are they consistent with or do they contradict other well-established explanations? How much of the phenomenon is explained? Can the explanation be used to explain or predict similar phenomena? How complicated or simple is it? What tests could support or undermine it?

When we apply these criteria to the scientific evidence and observations used to explain the ability of humans to make moral judgments, we find that they are the best and most likely explanation. They use well-established natural elements of brains and genes and studies of human and animal behavior that all play a role in our moral behavior. They are consistent with evolutionary, psychological, and neurocognitive theories that explain other related behavioral phenomena. There are no internal contradictions in these scientific explanations and almost all the behavior in question is adequately explained. Using behavioral techniques it is possible to predict behaviors with a reasonable degree of accuracy. Behavioral genetics and neurocognitive scientists are constantly testing and improving the natural explanation.

When you apply these criteria to Collins's supernatural explanation, you do not get a satisfying result. To explain natural phenomena, he uses an unnatural supernatural element. In fact his solution, God did it, does not explain the moral sense at all—it explains it away. It is the same as explaining the Moral Law as a magic spell. A rainbow is a real phenomenon, which is explained as reflected light being broken up by droplets into its spectrum of wavelengths. To say a rainbow is a divine sign by God to confirm that God will never again annihilate humans by flooding the entire world is not an explanation but a way to avoid an explanation.

Finally, even if you agreed with Collins after surveying all the evidence developed for a natural explanation of the Law of Good and Evil, it would be irrational and unscientific to leap to the conclusion "God did it." Collins himself cautions, "Faith that places God in the gaps of current understanding about the natural world may be headed for crisis if advances in science subsequently fill those gaps. Faced with incomplete understanding of the natural world, believers should be cautious about invoking the divine in areas of current mystery lest they build an unnecessary theological argument that is doomed to later destruction."[38] Ignoring his own advice, faced with a gap, Collins leapt.

## NOTES

1. C. S. Lewis, *Mere Christianity* (New York: HarperCollins, 2001), p. 5.
2. F. Collins, *The Language of God* (New York: Simon & Schuster, 2006), p. 24.
3. Ibid.
4. Ibid.
5. K. Heinrich and J. Spranger, *The Malleus Maleficarum* (New York: Dover, 1971).
6. A. Kors and E. Peters, *Witchcraft in Europe 1000–1700: A Documented History* (Philadelphia: University of Pennsylvania Press, 1972).
7. John Mark Ministries, http://jmm.aaa.net.au/articles/13127.htm.
8. Collins, *Language of God*, p. 24.
9. Ibid., p. 29.
10. C. Darwin, *Descent of Man* (New York: Penguin Classics, 2004), p. 146.

11. Collins, *Language of God*, p. 25.

12. C. Hitchens, *The Missionary Position: Mother Teresa in Theory and Practice* (London: Verso, 1997).

13. B. Kolodiejchuk, ed., *Mother Teresa: Come Be My Light* (New York: Doubleday, 2007).

14. F. Warneken, B. Hare, A. Melis, D. Hanus, and M. Tomasello, "Spontaneous Altruism by Chimpanzees and Young Children," *PLoS Biology* 5, no. 7 (July 2007), http://biology.plosjournals.org/perlserv/?request=get_documentsdoi-=10.137/journal.pbio.0050184; F. Warneken and M. Tomasello, "Altruistic Helping in Human and Young Chimpanzees," *Science* 311 (March 3, 2006): 1301–303; D. Hamer, "Rethinking Behavior Genetics," *Science* 298 (October 4, 2002): 71–72.

15. R. Bachner-Melman et al., "Dopaminergic Polymorphisms Associated with Self-Report Measures of Human Altruism: A Fresh Phenotype for the Dopamine D4 Receptor," *Molecular Psychiatry* 10, no. 4 (April 2005): 333–35, http://www.nature.com/mp/journal/v10/n4/full/4001635a.html.

16. Ibid.

17. Ibid.

18. A. Knafo et al., "Individual Differences in Allocation of Funds in the Dictator Game Associated with Length of the Argonine Vasopressin 1a Receptor RS3 Promoter Region and Correlation between RS3 Length and Hippocampal mRNA," *Genes, Brain, and Behavior* 7, no. 3 (April 2008): 266–75.

19. D. P. Barash, *The Survival Game: How Game Theory Explains the Biology of Cooperation and Competition* (New York: Times Books, 2003).

20. D. Knoch, A. Pascual-Leone, K. Meyer, V. Treyer, and E. Fehr, "Diminishing Reciprocal Fairness by Disrupting the Right Prefrontal Cortex," *Science* 314 (November 3, 2006): 829–32.

21. N. Stone, L. Cosmides, J. Tooby, N. Kroll, and R. T. Knight, "Selective Impairment of Reasoning about Social Exchange in a Patient with Bilateral Limbic System Damage," *Proceedings of the National Academy of Sciences* 99, no. 17 (August 12, 2002): 11531–36.

22. Collins, *Language of God*, p. 24.

23. Ibid., p. 263.

24. Human Genome Project Information: Behavioral Genetics, http://www.orn/gov/sci/techresources/HumanGenome/elsi/behavior.shtml. For a more detailed discussion, see the American Association of Science text "Behavioral Genetics" by Catherine Baker, http//www.aaas.org/spp/bgenes/

intro.pdf, and "Genetics and Human Behavior: Nuffield Council on Bioethics," http://www.nuffieldbioethics.org/go/outwork/behavioralgenetics/purlication_311.htm.

25. Collins, *Language of God*, p. 27.

26. J. Alcock, *The Triumph of Sociobiology* (New York: Oxford University Press, 2001).

27. Collins, *Language of God*, p. 27.

28. C. P. Van Schaik and C. H. Janson, eds., *Infanticide by Males* (New York: Cambridge University Press, 2000).

29. D. Farrelly, J. Lazarus, and G. Roberts, "Altruists Attract," *Evolutionary Psychology* 5 (2007): 313–29.

30. U. Segerstrale, *Defenders of Truth: The Battle for Science in the Sociobiology Debate and Beyond* (New York: Oxford University Press, 2000); E. O. Wilson, *Sociobiology: The New Synthesis*, fifteenth anniversary ed. (Cambridge, MA: Belknap Press, 2000); M. E. Borrello, "The Rise, Fall and Resurrection of Group Selection," *Endeavour* 29 (2005): 43–37; S. West, A. Griffin, and A. Gardner, "Social Semantics: Altruism, Cooperation, Mutualism, Strong Reciprocity and Group Selection," *Journal of Evolutionary Biology* 20 (March 2007): 415–32; D. S. Wilson and E. O. Wilson, "Rethinking the Theoretical Foundation of Sociobiology," *Quarterly Review of Biology* 82 (2008): 327–48.

31. Collins, *Language of God*, p. 28.

32. D. Lieberman, J. Tooby, and L. Cosmides, "Does Morality Have a Biological Basis? An Empirical Test of the Factors Governing Moral Sentiments Relating to Incest," *Proceedings of Royal Society London (Biological Sciences)* 207 (February 1, 2003): 819–26.

33. R. Trivers, "The Evolution of Reciprocal Altruism," *Quarterly Review of Biology* 46 (1971): 36–57; J. Clutton-Brock, "Punishment in Animal Societies," *Nature* 273 (1995): 209–26; S. T. Brosnan and F. de Wall, "Monkeys Reject Unequal Pay," *Nature* 425 (September 2003): 297–99.

34. M. Konner, *The Tangled Wing: Biological Constraints on the Human Spirit*, 2nd ed. (New York: Henry Holt, 2002); M. Ridley, *The Origins of Virtue* (New York: Penguin Group Books, 1998); R. Wright, *The Moral Animal* (New York: Vintage Books, 1994); Marc Hauser, *Moral Minds: How Nature Designed Our Universal Sense of Right and Wrong* (New York: Harper Collins, 2006).

35. Collins, *Language of God*, p. 23.

36. S. Pinker, *The Language Instinct* (New York: William Morrow, 1994).

37. P. Boyer, *Religion Explained: The Evolutionary Origins of Religious Thought* (New York: Basic Books, 2001); S. Atran, *In Gods We Trust: The Evolutionary Landscape of Religion* (New York: Oxford University Press, 2005).

38. Collins, *Language of God*, p. 93.

*Chapter 5*

# COSMOLOGY: ORIGINS OF THE UNIVERSE

> *Who established the course of the sun and stars? Through whom does the moon wax and wane? Who has upheld the earth from below and the heavens from falling? Who sustains the waters and plants? Who harnessed swift steeds to wind and clouds? What craftsman created light and darkness? Through whom exist dawn, noon, and eve?*
>
> —Zoroaster Yasna 44:3–6 (ca. 700 BCE)

## WHAT CAME BEFORE THE BIG BANG?

Collins accepts that the "universe began at a single moment, commonly now referred to as the 'big bang' some 14 billion years ago."[1] This moment is called a "singularity" by scientists and could be described by philosophers as the uncaused first cause of the universe. If you were a theologian or Collins, you might call it God.

When Collins asks the question, "What came before the Big Bang?" he implies it can only be God.[2] In *A Brief History of Time*, theoretical physicist Stephen Hawking describes the singularity as the beginning of time when the "density of the universe and curvature of space would have been infinite. This means that even if there were events before the big bang, one could not use them to determine what would happen afterward because predictability would break down at the Big Bang."[3] We cannot know what happened, if anything, before the big bang. If you accept the big bang as the beginning of space-time, the answer could be nothing. The Christian answer is that God changed nothing into something. This presents a problem. It assumes that something, the universe, is better than nothing, no universe. Is this assumption justified? If before the singularity only God existed, and if God were an absolutely perfect being requiring nothing, there is no rational explanation why God would create anything. Ask the question: if you were God, existing in timeless perfection, what would you do? Spark off a singularity that started the space-time continuum and wait billions of years for one planet to form around an undistinguished star in one of billions of galaxies? Then, using plate tectonics, fashion a strip of land called Palestine so when human beings evolved millions of years later, you could send your son there to save humans from their sins and join you in the "nothingness" of heaven? My question is, if this perfect being exists, how could he have created in such a slow, wasteful manner such an imperfect universe? Couldn't you imagine that an all-powerful God would devise a more intelligent and efficient process? The existence of an imperfect universe is incompatible with the claim that a perfect creator God exists.

But in Collins's opinion, "The Big Bang cries out for a divine explanation."[4] Is any explanation needed? If so, why does it have to be "divine"? How would Collins respond to the scientists' question, "What happened before God?" Why does the big bang demand an explanation and the divine does not? We have no way of knowing whether the big bang itself was a unique, uncaused event or was the result of a prior unique, uncaused cause, namely, God. Logic does not force either conclusion. In the absence of evidence we are dealing with opinions, not knowledge.

Collins also has difficulty relating God to time. "Only a supernatural force that is outside of space and time could have done that."[5] Okay, God is timeless, outside of time, and time becomes one of God's creations. However, Collins later argues, "If He [God] is not limited by time, then He is in the past present and future."[6] So now

God is in time and exists in eternity and we cannot think of a point in God's existence when God created time. Does God have a past before the big bang that created time?

Most Christian beliefs seem to require that God existed and acted in time. For example, God created the universe and the Garden of Eden with sinless humans. Then they subsequently changed the nature of the world by sinning. This original sin in turn required, sometime later, God to send his son, Jesus, to change conditions so humans could be free of the burden of original sin. After that Jesus was crucified and rose from the dead. Didn't these events take place in time?

Does God deceive us with the illusion of time? If this attempt to reconcile the notion of God with time as we experience it makes your head spin, you have no reason to be embarrassed. The concept of time is both simple to experience and difficult to explain, but the concept of God is impossible to explain or comprehend.

The physics of the big bang are extraordinarily complex because of the unimaginably colossal energy and mass involved. But most cosmologists are unwilling to abandon their attempts to understand the big bang, and they are not content to attribute it to another God of the gaps. I am not an expert in this arena, but I know that scientists constantly test and discuss natural explanations of the big bang. This area of science is evolving. If you are interested in exploring the scientific work in progress, there are popular explanations available.[7]

Collins continues to argue his point that God is responsible for the big bang by quoting the theoretical cosmologist/physicist Stephen Hawking: "If we find the answer to [why it is that we and the universe exist], it would be the ultimate triumph of human reason for then we would know the mind of God."[8] Collins uses this quote to imply that Hawking accepts the idea of a divine supernatural being or God. Clearly, if scientists come up with a unified theory of everything, it will generate discussions about its metaphysical implications, but it will not answer the "why" question or tell us the purpose for our existence. But Collins sees meaning that isn't there. All Hawking does is use the rhetorical phrase "knowing the mind of God" for effect. He does not imply that he believes in God nor that God has a mind, and he certainly says nothing about accepting a personal God like Jesus. Hawking is saying that it takes reason—human beings' most important mental faculty—to arrive at a natural and rational explanation of how the universe works.

The other Hawking quote Collins uses is, "We could . . . imagine that there is a set of laws that determines events completely for some supernatural being who could observe the present state of the universe without disturbing it."[9] This statement, if true, limits God's actions by making them completely dependent on natural law, not vice versa. This results in a completely determined universe with no free will and a deistic god that dispassionately and impotently observes the universe and humans without interacting with them.

Collins puts forward that "the universe began at a single moment," which he erroneously equates with an event Hawking and most physicists and cosmologists call "the singularity." However, in *A Brief History of Time* Hawking very clearly states that there is no proof that the universe has a beginning, end, or a creator. Science writer John Horgan gives a layman's interpretation of Hawking's "no-boundary proposal," stating Hawking's view of the big bang and what might or might not have preceded it:

> The no-boundary proposal addressed the age-old questions: What was there before the big bang? What exists beyond the borders of our universe? According to the no-boundary proposal, the entire history of the universe, all of space and all of time, forms a kind of four-dimensional sphere: space-time. Talking about the beginning or end of the universe is thus as meaningless as talking about the beginning or end of a sphere. Physics too, Hawking conjectured, might form a perfect, seamless whole after it is unified; there might be only one fully consistent unified theory capable of generating space-time, as we know it. God might not have had any choice in creating the universe. "What place, then, for a creator?" Hawking asked. There is *no* place, was his reply; a final theory would exclude

> God from the universe, and with him all mystery. Like his colleague, physicist Steven Weinberg, Hawking hoped to rout mysticism, vitalism, creationism from one of their last refuges, the origin of the universe.[10]

While I agree with Collins that *A Brief History of Time* was more purchased than read and understood, anyone who does read it will clearly see it is the description of a natural process for the generation and evolution of the cosmos. Hawking's use of the word "god," just like my use of the word, is not evidence of belief. Hawking is a man with a very creative, imaginative mind and has a lot of background knowledge. Though he is a competent mathematician, he usually collaborates with other mathematicians on the details. His real genius is in conceiving relationships and constructing theoretical propositions. He is a popular cult figure due to his controversial speculations about time travel, computers outthinking people, and so on. While he is coy and ironic about his beliefs, and by any reasonable definition, he is an atheist or at the very least an agnostic as illustrated by the following quotes:

> So long as the universe had a beginning we could suppose it had a creator. But if the universe is really completely self-contained, having no boundary or edge, it would have neither beginning nor end. It would simply be. What place then for a creator?[11]

> What I have done is to show that it is possible for the way the universe began to be determined by the laws of science. In that case, it would not be necessary to appeal to god to decide how the universe began. This doesn't prove there is not god, only that god is not necessary.[12]

> We are such insignificant creatures on a minor planet of a very average star in the outer suburb of one of a hundred billion galaxies. So it is difficult to believe in a god that would care about us or even notice our existence.[13]

When Hawking was asked by a reporter at a conference following the publication of his book if he believed in God, he was reported to have responded, "I do not believe in a personal God."[14] On the other hand, Hawking realizes the intense interest that believers have in using science to support their belief systems, so he includes speculations and references to God in his writings to promote interest in his books and lectures among nonscientists.

Even if there were no scientific natural explanation for the big bang, the metaphysical explanation would be that it is the final, ultimate uncaused cause of the universe. No eternal supernatural being called God is needed as a separate uncaused cause existing outside of the universe. Again, God is used to fill a hopefully temporary gap in scientific description of the origins of the universe.

## WHY IS THERE SOMETHING RATHER THAN NOTHING?

The fact that there is something—a material, natural, imperfect universe—is evidence against a divine, perfect, immaterial, supernatural God. If God existed, he would be the only being existing prior to creation. There would be only God's supernatural being and no inferior, unnecessary, imperfect natural beings. If God existed, and for some unfathomable reason wanted to create a world, God's perfect nature would demand that it be the best possible world, namely, a perfect world. But it is easy to see ours is not a perfect or even the best possible world. Therefore, God does not exist.[15]

Why is there something rather than nothing? The scientific naturalist answer is that matter and energy cannot be created or destroyed. Energy can be converted into matter ($E = mc^2$) and matter can be converted into energy, but the net amount of matter/energy in the universe is constant. Matter/energy could then be said to be eternal. It would be reasonable to conceive of the universe existing eternally without beginning or end, going through cycles of a big bang expansion, ultimate collapse in a "big crunch," and then creating a new singularity with a big bang, ad infinitum. This cyclical universe hypothesis is not the current scientific consensus, but it is more reasonable than supposing a supernatural creation.

## WHAT IS THE PURPOSE OF CREATION?

All believers must propose a divine purpose for creation. What did God intend to happen as a result of creation? Does God need the "fellowship" of humans, as Collins suggests? This seems more than implausible. Humans need the fellowship of other humans, but we don't seek the fellowship of things we create. We enjoy and use the things we design and create, our inventions, but we don't need to have fellowship with them. Why would God? Do you enjoy fellowship with your computer? Does your computer worship you or you it?

God's relationship as creator to creature is not one of equals. God, by his nature, does not *need* our worship, our prayers, or our love. If you insist on god having a purpose for the universe, it certainly would not have anything to do with the insignificant human life-forms that temporarily occupy the insignificant planet Earth spinning in the immensity of the total universe over the course of the billions of years of its existence.

However, I would like to question Collins's words, "If God exists and seeks to have fellowship with sentient beings like ourselves . . ."[16] This implies that God was unhappy alone in heaven prior to the big bang and somehow needed humans. Even if a perfect supernatural being had the imperfection of needing fellowship, God could have created supernatural beings closer to God's nature, like angels, to keep God company. This is like saying that God created the nose to hold up eyeglasses. This is not the purpose, or better, the function of the nose. While it serves to keep eyeglasses in place, this is only a coincidence, a consequence, which we cannot characterize as the primary purpose of the nose. Humans and the rest of the universe are the consequence of natural laws and are not here to make God happy. Based on their history, humans seem like a poor choice for worshippers or companions. Couldn't God have created a race of more intelligent, compassionate, cooperative beings than human beings for fellowship? The notion of fellowship does not follow from Collins's cosmological arguments for an impersonal creator God. Even if this creator God exists, it does not help Collins, who is defending the idea of Jesus, a personal God. The concept of a personal God is an issue that I discuss in chapter 9.

Next, Collins speculates on the rarity of life, especially intelligent life, in the universe to support the idea of humans being special to God. I agree with his conclusion that intelligent life elsewhere is probably rare, although not implausible. I also agree with his opinion that discovery of life, or even intelligent life, elsewhere in the universe would have little bearing on the basic argument about God's existence. Science would look for natural explanations for the extraterrestrial life and Collins would argue that all things possible are possible for God and maybe God has fellowship with the aliens as well. The probability that we will make contact with aliens is so low as to make their existence for all practical purposes irrelevant to our debate.

Asking the question "What is the purpose of the universe?" is based on two unproven assumptions. First, you are assuming everything has a purpose, and second, you are assuming the purpose was assigned by someone. One good, rational answer to "What is the purpose of the universe?" is that it has no good purpose other than what we humans attribute to it. Purpose is a creation of the evolved human mind, a consequence of how it operates. That may not be an answer you like, but your opinion does not alter the truth of the concept nor the evidence supporting it. All purposes and meanings for the universe are products of the human mind. As far as we know only a few higher animals have developed the capacity for conscious, intentional, goal-oriented actions. You cannot reason by analogy that if we humans have needs and purposes, so does God (see chapter 9). It is implied that if God did have a purpose for the universe and humanity, then God has the power to ensure that this purpose would be achieved. God's intentions cannot be frustrated. The outcome is known and guaranteed. If this is true, the world is deterministic and there is no place for free will.

## ANTHROPIC COINCIDENCES

As evidence for the existence of an intelligent creator responsible for the universe, Collins next turns to the *anthropic principle*, which he describes as "the idea that our universe is uniquely tuned to give rise to humans."[17] Copious literature states, restates, and reformulates the anthropic principle, but Collins's version is a common variation. The anthropic principle, which could more properly be called the anthropic *coincidences*, is based on the numerical value of measurements and calculations that predict or describe certain relationships in nature. These mathematical constants, which sustain the laws of physics, have just the right value needed to support life. Since scientists can give no reason why the constants have these critical values, Collins believes there must be a supernatural explanation, namely, the God of the gaps. Paraphrasing Collins, "God so loved humans that he fine-tuned these constants just right so that after about 14 billion years humans would appear on the planet Earth." But are the constants really fine-tuned to make life possible or to make it necessary? Do they permit carbon-based life or do they require the formation of intelligent life? There is nothing about the constants themselves that would make the evolution of intelligent life a necessary outcome. If a meteor hadn't knocked off the dinosaurs, intelligent mammals would have had a hard time replacing them.

To explain how Collins and others illogically use this concept to prove the existence of God, I would like to borrow an incident described by Michio Kaku in his book *Parallel Worlds*.[18] As a child, Kaku remembers his teacher stating, "God so loved the earth that he put the earth just right from the sun." Scientists and philosophers often refer to this as the "Goldilocks Zone." Just like the porridge in the fairy tale, the Earth was positioned with respect to the sun so as to be not too close and too hot and not too far and too cold, but just right to support intelligent life. The big bang expansion was not too fast to disintegrate into tiny pieces or too slow to collapse into a point of pure energy, but just right to allow stars and planets to form. We have the picture of God at the cosmic control panel manipulating the distance to the sun to ensure the survival of his intelligent, loving creatures. What Kaku's teacher did not mention was that this distance is constantly changing. Early in the formation of the cosmos, the Earth was so hot that no life could survive. The future expansion of the sun will again wipe out all life. In terms of cosmic time, we are simply enjoying a brief interlude in a much longer cosmic drama.

Collins enumerates then dismisses two of three possible responses that scientists might use to explain how these physical constants were aligned so precisely without a god to arrange it. First, there may be a large or infinite number of universes or multiverses and, simply by statistical probability, one or more would have physical constants that would lead to the creation of intelligent life during the universe's evolution. Theoretical physicists, mathematicians, and cosmologists are continually developing and updating the multiverse hypothesis. Kaku's book and Max Tegmark's cover story in the May 2003 issue of *Scientific American* describe in layman's language the evidence for multiple universes.[19]

Cosmologist Edward Tryon theorized that quantum fluctuations, which exist even in a void (vacuum), could create matter. Hawking has proposed that singularities are popping up all around us like bubbles in a large pot of boiling water. Most are tiny and have physical constants that lead to their almost immediate destruction. Some grow bigger, forming black holes in the middle of galaxies or in space. Some may develop into alternative universes. One of these initial conditions generated the physical constants that produced our universe with intelligent life. Without intelligent life to observe them, alternative universes would be completely unknown. For all we know, there could have been uncountable empty and unstable universes before ours. Such a natural explanation, while not yet confirmed by evidence, is at least plausible.

Collins finds this possibility "logically defensible, but this near infinite number of unobservable universes strains credulity."[20] Oddly, this credulity gap in the multiverse argument is too far for Collins to leap over, but Bible stories—the virgin birth, the resurrection of the dead, and numerous miracles—are all credible. While I do agree with Collins that the existence and nature of multiple and/or parallel universes is counterintuitive, it is a scientific question awaiting a convincing answer and cannot be dismissed as a reasonable alternative natural explanation.

Collins's second option is that the universe "just happened to have all the right characteristics to give rise to intelligent life."[21] Collins rejects this option on the grounds that the odds against such a universe developing after the big bang are "enormous." One reason the odds are "enormous" is that, in calculating the odds, they are exaggerated by combining the probabilities of several independent constants. To illustrate, what are the odds against the possibility that you are now sitting in a chair? Rather high. What are the odds against the chair being in an airplane? Somewhat higher. What are the odds against the chair being purple and the only chair in the airplane, which is over the geographic center of Alaska, has six engines, and the pilot is a Chinese-speaking, deaf, Australian aborigine, who is blind in the left eye and has an amputated right big toe . . . you get the idea. Whatever the odds against any one independent constant having a particular value, the odds against multiple constants having particular values are "enormous."

He again quotes Stephen Hawking's metaphysical message as indirectly supportive of his intuitions. But Hawking said, "This work on inflationary models showed that the present state of the universe could have arisen from quite a large number of different initial configurations. This is important because it shows that the initial state of the part of the universe that we inhabit did not have to be chosen with great care."[22] Not so unique and not so fine-tuned. Hawking also said,

> The weak Anthropic Principle states that in a universe that is large or infinite in space and/or time, the conditions necessary for the development of intelligent life will be met only in certain regions that are limited in space and time. The intelligent beings in these regions should therefore not be surprised if they observe that their locality in the universe satisfies the conditions necessary for their existence. It is like a rich person living in a wealthy neighborhood not seeing any poverty.[23]
>
> We may be surprised that conditions in the universe happen to be suitable for life, but this is not evidence that the universe was specifically designed to allow for life. We could claim god ordered the physical constants, but it would be an impersonal god. There is not much personal about the laws of physics.[24]

As stated by Collins, this explanation depends on the chance occurrence of an improbable collection of fifteen (or however many you want to use) constants. But just listing constants is a description of nature as we find it, not an explanation. The probability that we live in a universe that gives rise to intelligent life, given that the universe has certain characteristics that support intelligent life, is not close to zero; actually, it is one, or 100 percent. So Collins seems to be saying that option two, chance, can be dismissed as "straining credulity." But in the absence of a scientific understanding of the mechanisms responsible for the constants, it seems rational to assume they are the result of either chance occurrences or a yet-to-be-discovered unifying law. If, as Stephen Hawking suggests will happen in the next twenty years, scientists are able to come up with a compelling and convincing unifying theory that demonstrates the natural causes and relationship of the constants, Collins's argument collapses.

Collins uses an analogy offered by the Canadian philosopher John Leslie to illustrate how these constants were not the result of blind chance but were "fine-tuned" in an intentional way by the intelligent designer at his master control board in the sky. When fifty expert marksmen assembled as a firing squad fire at a victim and the victim walks away unscathed, the only plausible explanation is they all purposefully missed. The only other alternative given by Collins is chance. However, there are additional explanations. Suppose all fifty rifles came from the same defective manufacturer and all fired far to the left of the sighted target. It could be that a "theory of everything" will reveal all these constants have their value determined by a single constant on which they depended. There could be a defective batch of ammunition. The fifty marksmen might have ingested a toxin that affected their eyesight. Or, there could be a combination of these so that the end result was fifty misses. Intentional missing is not the only explanation with a reasonable possibility. In any event, the firing squad, a single isolated event, is not an apt analogy for a dynamic evolutionary cosmic process.

I would like to suggest another analogy. What are the odds that you would be born with your unique set of genes, called your genome? The odds are 1:1, 100 percent. Now, your genome came from the 30,000-odd genes you got from your mother when matched with the 30,000-odd genes you got from your father. This was a chance event. If you changed even one gene, the results would be a different genome than the one you have. Your mother's and father's genomes came from a combination of the contributions of the 120,000 genes of your paternal and maternal grandparents. This process goes back generation after generation. You go back a long way since you must have had distant relatives who date to the separation of the species *Homo sapiens*. Your two parents become four grandparents and eight great-grandparents and sixteen great-great-grandparents. After twenty generations, you would have $2^{20}$, or 1,048,576, distant relatives who contributed to your genome. If, as Collins and I agree, *Homo sapiens* has been around for about 195,000 years, and an average generation was approximately 40 years, you could go back 4,875 generations. The probability of starting with the original group of great-great-great . . . grandparents, and considering the chance of all the right matings and the right ova maturing and the right sperm from the millions injected fertilizing it and the right mutation at the right time, that you would end up with your genome and not with a million other possible genomes is astronomical. Yet whatever the odds, it is a fact—the result of entirely natural events, and we don't need to assume that there was a supernatural fine-tuner who miraculously designed and put together your genome.

Before an event, calculation of the probability of one specific outcome among many possible outcomes gives us a measure of the likelihood it will occur. After an event, there is only one actual outcome. There are no longer any possible outcomes. Probability is only related to the future. The facts and events that have happened all have a probability of 100 percent. The probability of life existing in our universe is 100 percent.

The probability of life existing in a future universe or an imagined universe can be calculated, and the probability of another life-containing universe can be estimated. However, these calculations are irrelevant to whether our life-containing universe exists. Just because a result is highly improbable does not mean that it cannot be due to chance and requires a special explanation. If you took a million decks of cards, shuffled each deck seven times (the number of times required to assure random distribution), and dealt them to find that every deck was in perfect order from ace to king and in suit order—diamonds, hearts, clubs, and spades—the event would be a real natural occurrence no matter how improbable such a result would be before you started the process. After the event occurred, it would be history. The fact that a repeat performance is highly unlikely does not change the fact that it happened. No matter how astronomically high the odds against this result, it is not impossible and after it has occurred its improbability is irrelevant. Nothing supernatural would be needed to explain the result.

Finally, we come to option three: "The precise tuning of all of the physical constants and physical laws . . . reflects the action of the one who created the universe in the first place."[25] For Collins this option "provides an interesting argument in favor of a creator," while he admits "no scientific observation can reach the level of absolute proof of the existence of God."[26] This assumes that, since we have no universally accepted natural answer to this assumed problem of the constants, the only explanation is that there is some supernatural intelligent being capable of having purposes for its actions who is responsible. This accepts as the only explanation the belief in the supernatural, which is exactly what Collins is trying to prove. This is the classic God of the gaps.

The explanation that an incomprehensible God "did it" is no explanation at all. It is like saying it's magic, which explains nothing. Alternative explanations are possible and more probable.[27] The first is that there is no other intelligent being selecting the constants. It is intelligent humans who are hardwired to have reasons and purposes, who imagine this other external intelligent being (God of the gaps), when, in fact, there is no such being. The universe does not need to have a grand ultimate purpose. The universe could be a completely determined chain of caused events without any external grand purpose, or goal, or final product at all.

Further, we might ask, can we determine the creator's (intelligent designer's) purpose by examining the product of creation? Was humanity the ultimate purpose of the creative process? The fact that 96 percent of the universe consists of empty space, dark energy, and dark matter that seems to serve no intelligible purpose, and that rather than taking six days, it took 13 billion years of creating to get to the big event, that is, humanity, does not seem to support this interpretation. If you look at the universe as a whole, it would seem that the creator wanted to create stars, not intelligent life. God created trillions of them. Do you think humans, who inhabit one planet (Earth) orbiting one star (the sun), are just an insignificant by-product of the major work of creation, or are they the creator's primary purpose? Not only is the Earth not the center of the solar system but also our galaxy, the Milky Way, is not even the center of our universe. There are billions and billions of stars, which have "life spans" of billions of years. All human beings, who ever lived or are likely to live, have very short lives. On the cosmic time scale, we seem to be an afterthought. Humans are only a tiny fraction of the mass and energy in the universe, a drop in the bucket in terms of the total content of the universe.

Was the creator fine-tuning God's creation computer in the sky with the purpose of creating human life in one tiny point in the universe? What life-form would the physical constants that God selected prefer? Which life-form would be most numerous and long lasting? Bacteria! Bacteria are the longest-lasting and most numerous form of life on earth. There are an estimated 30 million species of bacteria and billions of individuals in each species. Bacteria were around millions of years before any humans and will still be around when the approaching sun has killed all humans. They can live in environments and under conditions that no human could survive. Ten percent of your body weight is bacteria. The intelligent designer may have intended humans primarily to serve as just another transporting vehicle and source of food for bacteria.

Even insects are more numerous and more environmentally robust than humans. There are more than 250,000 species of beetles alone. A visiting intelligent space alien, observing our universe, would think it is better designed to support germs and bugs than humans. If the alien happened to visit before humans evolved or after we have become extinct, this would certainly be its conclusion.

It could be argued that the anthropic principle is an argument against the existence of a creator desirous of creating intelligent life. Look at the universe. As C. S. Lewis observed: "If we used that as our only clue, then I think we should have to conclude that He was a great artist (for the universe is a very beautiful place), but also that He is quite merciless and no friend to man (for the universe is a very dangerous and terrifying place). The universe is full of fiery stars and colorful planets, and all sorts of awe-inspiring, beautiful demonstrations of destructive energy."[28] The universe is indeed a very hostile place for life, which is why species constantly go extinct and constantly struggle to survive.

There are two senses in which we ask the question "Why?" First, you can ask what caused an event, for example, "Why did Lincoln die?" Because John Wilkes Booth shot him. Or you can ask for a reason or purpose for an event, for example, "Why did Lincoln die?" To avenge the defeat of the South and change the government. You can ask the first question about any event in nature, but the second can only be asked if you assume the event was caused intentionally by a being with purposes or desired results. The only such intentional being known to exist that infers that actions and events have purposes or meaning in the second sense of "why" is the human being. Christians believe there is another such being, God. The naturalist view of the universe is that there is no evidence for intention or purpose imposed on the universe by a separate supernatural being. In our natural universe, man is the measurer and the measure of all things. Only when conscious intentional life evolved could man begin defining his own purposes, imposing his own meanings, and creating gods in his own image.

Suppose God wanted to create intelligent life for fellowship, as Collins believes. Could God have used different physical constants and still have ended up with a universe containing humans? If your answer is no, then these constants are unique and cannot be changed without making evolution of human life impossible. God had to use this one unique set of constants. Aren't you then saying that God's power is limited to obeying natural law, which he does not control? On the other hand, if your answer is yes, then God is all powerful and there is no reason he could not produce humans with any constants he chooses. But that would mean that these constants are not really unique and very fine-tuned. These constants are only one of several different sets of constants that could produce life, so there is no need to explain why only one improbable set of constants is absolutely necessary and eliminates the argument for a fine-tuner.

We have no evidence of life, intelligent or otherwise, elsewhere in our universe. While it may, and probably does, exist in some distant galaxy or alternative universe, it certainly is rare enough to conclude that creation of life, as important as that is to us, is not the intention behind the "creation" of the universe.

Consider the significance of intelligent life in the really grand sweep of galactic time. Intelligent life on Earth will disappear from the universe in the next 150 billion years. The cosmos is evolving and the universe is expanding. If we manage to avoid nuclear annihilation, a natural megacatastrophe, or a cataclysmic collision with a massive object from space, we will die from the big freeze as Earth's temperature goes to absolute zero and molecular motion stops. This will put an end to any life in the universe. So what kind of impersonal creator do we have that "precisely tunes" the physical constants and laws of nature to achieve this result? In the cosmic time scale of creation, the existence of humans, saints and sinners, will have been a tiny sideshow, not the crowning achievement of Collins's loving God who died on the cross and rose again to reward or punish us inconsequential humans for our brief period of misbehavior.

In his book, Collins inserts a brief section on quantum mechanics and the uncertainty principle, which covers microscopic events that are totally random and unpredictable by any natural law. Examples of such events include the inability to accurately determine both the position and the velocity of a subatomic particle at the same instant of time and the rate of decay of radioactive uranium-238. This does not appear to be an argument for an intelligent, rational designer, since actions governed purely by chance are uncontrolled and unplanned and not designed.

If our actions are ultimately unpredictable and due to totally random quantum events, then free will becomes an illusion. We would have to abandon the concepts of praise and blame for our random behavior. The implications of quantum theory and the role of randomness in explaining the universe is still a subject of intense debate in the scientific community. Using the mathematics of quantum theory produces highly predictable results. When reality gets reduced to fundamental particles and forces, common sense, free will, and causality encounter problems. But I don't see how it helps Collins's argument to conclude that, in Einstein's words, the creator "plays dice" with his creatures. These are uncaused events that occur at the submicroscopic level. The causal laws of classical mechanics are still valid for what happens in the macroscopic human-sized world. As a practical matter, our worldview needs to be consistent with the real human-sized world if it is to be really relevant.

For Collins and other scientific thinkers including myself, "the scientific worldview is not entirely sufficient to answer all the interesting questions about the origin of the universe."[29] Answering *all* the interesting questions about the universe is an impossibly high standard, but surely science has the best answers to date. Does any religion provide satisfying answers to all the interesting questions about the origin of the universe? The recurring answer that an incomprehensible God did it is an answer that explains nothing. It's like the answer, "It's magic."

What about the gaps in our scientific knowledge? Collins offers us the God of the gaps as the source of the big bang and the physical laws that led to intelligent life on Earth. But in this case, the God of the gaps is the impersonal god of the deists like Einstein and possibly Hawking. Collins proposes a theoretical supernatural being behind a natural system of laws. I believe there is an inherent conflict between a worldview that includes the supernatural and a worldview that excludes it. Natural explanations are sufficient and intellectually satisfying even though they are incomplete. Collins seems to believe the inclusion of a supernatural world is necessary and gives an outline of assumptions and consequences involved to achieve a "rational" synthesis of a Christian worldview that I will address in chapter 8.

## WHAT ABOUT GENESIS?

As part of this argument, Collins points out that the two creation myths of Genesis 1 and 2 are entirely compatible with the concept of the big bang if you consider these books poetic literature. He also states that sacred "texts that seem to describe historical events should be interpreted as allegory only if strong evidence requires it."[30] He seems to be saying that if scientific evidence conflicts with the Bible, the science is accepted and the Bible should be reinterpreted to fit the new facts. Once he decides to treat Genesis as a poetic story, he finds it "compatible with scientific knowledge of the Big Bang."[31] In his opinion, *compatible* means that this is one acceptable interpretation of the story but does not mean it is the only or the true interpretation. The big bang can also be found to be compatible with many other and very different creation stories or poetic myths. Genesis 1 and 2 are stories that could be reinterpreted to fit a variety of theories or facts. But they are clearly not science. The specifics of Collins's allegorical interpretation are discussed chapter 6.

Even if we accepted Collins's assumption of a divine cause for the big bang, and even if there were a divine fine-tuner to ensure intelligent life in the universe, his assumptions would only support the philosophers' god—an impersonal spiritual force; a first uncaused cause; a supernatural, intelligent creator—not a personal God like Jesus. Collins's case for a personal God remains unproven and unlikely.

## NOTES

1. F. Collins, *The Language of God* (New York: Simon & Schuster, 2006), p. 64.
2. Ibid., p. 66.
3. S. Hawking, *A Brief History of Time* (New York: Bantam Press, 1998), p. 46.
4. Collins, *Language of God*, p. 67.
5. Ibid.
6. Ibid., p. 81.
7. B. R. Greene, *The Elegant Universe* (New York: W. W. Norton, 1999); V. J. Stenger, *Has Science Found God?* (Amherst, NY: Prometheus Books, 2003); M. Kaku, *Parallel Worlds* (New York: Doubleday, 2005).
8. Collins, *Language of God*, p. 62.
9. Ibid., p. 80.

10. J. Horgan, *The End of Science* (New York: Helix Books, 1995), p. 94.
11. Hawking, *A Brief History of Time*, p. 141.
12. Interview in *Der Spiegel*, October 17, 1988.
13. BBC TV Production, "Master of the Universe," 1990.
14. Mark O'Brien, Pacific News Service, http://www.pacificnews.org/marko/hawking.html.
15. R. R. LaCroix, "Unjustified Evil and God's Choice," in *The Impossibility of God*, ed. M. Martin and R. Monnie (Amherst, NY: Prometheus Books, 2003), p. 116; L. Resnick, "God and the Best Possible World," in *The Impossibility of God*, ed. M. Martin and R. Monnie (Amherst, NY: Prometheus Books, 2003), p. 274.
16. Collins, *Language of God*, p. 71.
17. Ibid., p. 74.
18. Kaku, *Parallel Worlds*.
19. M. Tegmark, "Parallel Universes," *Scientific American*, May 2003, pp. 40–51.
20. Collins, *Language of God*, p. 76.
21. Ibid., p. 75.
22. Hawking, *A Brief History of Time*, p. 132.
23. Ibid., p. 124.
24. G. Benford, "Leaping the Abyss," *Reason*, April 2002.
25. Collins, *Language of God*, p. 75.
26. Ibid., p. 78.
27. V. Stenger, *God: The Failed Hypothesis* (Amherst, NY: Prometheus Books, 2007).
28. C. S. Lewis, *Mere Christianity* (New York: HarperCollins, 2001), p. 29.
29. Collins, *Language of God*, p. 80.
30. Ibid., p. 83.
31. Ibid., p. 150.

# *Chapter 6*
# THE BIBLE

*[The Bible] has noble poetry in it; and some clever fables; and some blood drenched history; and a wealth of obscenity; and upwards of a thousand lies.*

—Mark Twain

As Collins describes his journey from atheism to theism and then to Christianity, he cites the Bible as evidence of God's nature and intentions. Collins primarily if not exclusively uses the New Testament. While he speaks of reading "biblical and non-biblical accounts," he does not seem aware of the writings of scholars who have devoted lifetimes to examining, dating, translating, and analyzing the authorship of the varied collection of texts that were assembled into the book we call the Bible. The Bible cannot be relied upon as a historically accurate detailed description of events, especially miraculous events.

## MY EXPERIENCES WITH THE BIBLE

The Bible was not a central part of my Catholic upbringing, which depended on rituals and sermons to bind the faithful to the hierarchy. The analogy my teachers used was that the Bible is like the Constitution. An authoritative body is needed to interpret it. For the Constitution, this interpretive body is the Supreme Court; for the Bible, it is the Catholic Church. We were taught that the kind of individual interpretation espoused by Protestants had produced denominational anarchy and had infested the Gospel with ever-growing errors and disputes. Using individual readings to interpret the "real" meaning of Bible passages has spawned a profusion of theologians, each with their own version of Christology (the study of the nature of Jesus Christ and the religion he taught), theology (the nature of the divine), and apologetics (ways to defend a set of beliefs). This is "cafeteria Christianity," where you select only what appeals to you.

Nevertheless, most of the Bible's authors were skillful writers who crafted gripping stories that maintained my interest (except for all those "begat" sections). One section of the Bible that gave me pause was when Yahweh orders Abraham to sacrifice his son. At the time, I believed this story described a real event. I already wondered why so many animals had to be killed to please God. That was bad enough, but asking a father to kill his beloved son, Isaac, raised questions of God's goodness in my mind that I could not entirely resolve. These questions could not be resolved even by knowing that a sheep would be substituted at the last minute. What about all the misery God was causing Abraham? Consider that Isaac was his favored and much-desired heir by Sarah, his barren wife. Abraham had already sent Ishmael, his other son by his slave Hagar, into the desert to die to preserve Isaac's inheritance. Killing Isaac would have destroyed Abraham's whole purpose in living.

Some believers see the message in this passage as a demand by God for absolute, unquestioning obedience, even if it requires loss of all human happiness. Some defend it as a sign from Yahweh that human sacrifice was no longer needed to appease God and, henceforth, animal sacrifice would be sufficient. Theologians have multiple interpretations of this story. None of these explanations helped me retain my belief in a loving God deserving of worship.

The adventures of Noah's Ark, Joseph and his brothers, Moses' travails in the desert, the schemes of David, and the story of Job are other page-turners. Some scholars have identified these stories as rewrites of older stories from neighboring cultures. As literature, the Bible stories make for a thought-provoking, emotional, roller-coaster read.

If you are at all interested in a reasonable and plausible answer to the question of whether a personal God exists, you must read the Bible from cover to cover before you decide whether it can be used as evidence. I agree with Isaac Asimov, a great scientist and writer, when he said, "Properly read, the Bible is the most potent force for atheism ever conceived."

Read both the Jewish Bible, which Christians call the Old Testament, and the New Testament. If they are the word of God or an inspired message from God, every word is important. Why would God insert or allow the authors or copiers of God's message to insert anything that was not of transcendental importance and divine concern? God wants you to know the family trees of the twelve tribes of Israel, how and when to sacrifice animals, and so on. So, don't skip a word. Even if you come to the conclusion that it is just a book, the time will be well spent. It contains some great quotations, ideas, poems, and morality lessons.

Recent studies show that Americans not only have low scientific literacy but are also biblically and religiously illiterate.[1] Although 92 percent of American households have one or more Bibles, only 59 percent read one even occasionally. Few have more than a superficial knowledge of it. About 50 percent could name one of the New Testament Gospels, and only a third could name all four. Only 42 percent know Jesus delivered the Sermon on the Mount or could name five of the Ten Commandments. Yet, strangely, if you ask the average Christian to name the most important book written, they will respond "the Bible." They know stories in the Bible, but they do not know the story of the Bible. So I am going to discuss the Bible based on the assumption that many readers may not be all too familiar with its history and contents.

## ANALYSIS OF THE BIBLE

Let's take a look at your Bible. What do you see? It looks like any other book. When you open it, you find it is written in English. This is obviously a translation, for none of the authors or Bible characters could speak or write this modern language. If the original writings were divinely inspired and contain the true stories and messages that

God wanted humans to have, then you need to be sure the translation is accurate.

The first English translation from a Latin Bible was by John Wycliffe in 1382. In it Jesus and Moses have old English words such as "thy" and "thee" and "giveth" and "dwelleth" put in their mouths. These may have been revised into modern English in your Bible, but they are still not the original words uttered by biblical figures. If you lived before 1382, you would be reading a Bible in Latin. This was not the language of Moses or Jesus either.

If you were a believer before Latin texts were available in 393 CE, and if you were one of the fortunate few to have owned something as rare as one or more of the books of the Bible, your Bible texts would have been written in Greek. This was not the language of Moses or Jesus, but it is the language used in the oldest manuscripts of the New Testament. The first time the various books were collected in a single volume was probably the Bible commissioned and financed by Constantine in 330 CE.

The Jewish scriptures, which were included when the Bible was put together, were in Hebrew except for parts of Ezra, Jeremiah, and Daniel, which were written in Aramaic. Hebrew, or the Aramaic version, was the common spoken language in Palestine at the time of Jesus. In our quest to get the true, unchanged messages in the Bible, we need to look at the original manuscripts. But there are none. The oldest copies we have were produced long after the originals were written. So, we really cannot be sure how much the details of the stories and sayings in the copies that survived really record what was originally written.

These early manuscripts were hand copied from oral accounts and previous writings by many scribes in different locations and included both intentional and unintentional addition, and omissions. And what do we know about the original authors? Not much. During biblical times it was common practice for writers, in order to promote the acceptance of their writings, to attribute them to other authoritative persons or trusted leaders.

This makes it impossible to determine who wrote Matthew, Mark, Luke, and John, but we do know it could not have been the uneducated, probably illiterate, apostles. Few of the apostles were alive when the first Gospels were written. The Gospels were not assigned names until fifty or sixty years after they were composed. We cannot answer the question of who wrote Genesis or Exodus and so on, but we know that whoever they were, they were not eyewitnesses to the events.

It is clear that the Bible is not a book like other books in your bookcase. It is not an original edition and its authors are unknown. When you open it and look at the table of contents there is another difference. As we have noted, it is divided into two sections called Old Testament and New Testament, which are very different from one another. The section called Old Testament or the Hebrew Bible contains the Jewish scriptures discussed by Jesus and was written hundreds of years before his birth. It appears to be an account of the Jewish people and their special covenant with their powerful, wrathful, and demanding protector God, Yahweh. The Old Testament details Yahweh's laws for the Jews, including the Ten Commandments. The New Testament is concerned with the life and teachings of Jesus and doesn't mention Yahweh at all. This is not a single historical narrative with a beginning, middle, and end, but a collection of disconnected narratives and other types of literature.

So, when you read the Bible, you are depending on these ancient scribes and translators and their modern equivalents for the accuracy of the details. The Bible can be studied as a historical document by trained scholars.[2] While most of us will have read at least some parts of the Bible, very few of us have made a careful scientific study of it. Bible scholars have to study Hebrew, Greek, Latin, Coptic, and other languages. They must be familiar with the history and archeology of the times. They must learn to date manuscripts and recognize writing styles.

Examining the Bible in this way uncovers facts that raise questions about its accuracy. The Old Testament was a transcription of oral history and stories of the Jewish people, mostly written in Hebrew. Hebrew has no vowels and twenty-two consonants, some of which have no English-language equivalent. Yahweh is written YHWH. Hebrew is written as a continuous series of letters from right to left with no capitalization, punctuation, or separation into sentences, paragraphs, or words. (The familiar chapter numbers were added to the Bible in 1228 by Stephen Langton, archbishop of Canterbury, and verse numbers were added in 1551.) So if you read a Hebrew text that said DYWSHTHDRTNYRFTNTHMT, this could mean "Do You WaSH THe DiRT oN YouR FooT iN THe MoaT?" or "Do You WiSH THe DaRT iN YouR FaT iN THe MeaT?" Most meanings can be derived from context and writing style, but there are still words that are ambiguous and words whose meanings are disputed by Bible scholars.

Additionally, the Hebrew language was changing during this period, adding to the difficulty of translation and interpretation. Understanding the meanings of words depends on knowing their ancient context. For example, "The Cardinals are in town" could mean the Roman Catholic clergy was meeting, or the St. Louis baseball team was playing, or a flock of red birds have nested, or Stanford University's football team was visiting—depending on the context.

In the second century CE, the Jewish Bible was translated into Greek for Jews living outside Palestine. This version is called the Septuagint. Greek has twenty-four letters, including seven vowels. The translation had to create new words to reflect the older Hebrew meanings and concepts.

The New Testament was originally written in Greek. The first written documents were the epistles or letters of Paul composed from 50 to 63 CE that comprise almost half of the twenty-seven books of the New Testament. The original letters were lost or destroyed, and the oldest copy available to biblical scholars to reconstruct the original is known as Papyrus 46, produced in Egypt in 200 CE. There is much debate about how many letters Paul wrote. Bible scholars have collected about eight hundred copies of letters attributed to Paul, most incomplete and none of them identical.

The first Gospel was attributed to the author called Mark and was written about 60 CE. The first time the name *Mark* was assigned as author is in 150 CE. It was the first attempt to tell a story about Jesus' mission. Mark's writings were available to and used by the authors subsequently called *Matthew* and *Luke*. Other unnamed authors, writing in about 85 CE, contributed material. These authors are identified by letters such as *Q*.

Because of this common source relationship, Mark, Matthew, and Luke are called the *synoptic Gospels*. The Acts of the Apostles (believed to be written by the same author as Luke) and the Gospel of John were written in about 100 to 120 CE. The oldest text copy of a Gospel is a scrap from John dated 125 CE. What did the missing links—namely, texts between Jesus' death, dated at 30–33 CE, and 125 CE—really say?

Bible scholars also point out that between 200,000 and 400,000 textual versions of the Bible in Hebrew, Aramaic, Greek, and Latin were copied, edited, and changed over the centuries before being used to assemble the Bible as we know it. Contrary to Collins's impressions, the first included texts—the letters of St. Paul—were written forty to fifty years after the death of Jesus. Paul never met Jesus nor was he an eyewitness to his miracles, crucifixion, burial, or resurrection. (The only appearance of Jesus to Paul was an auditory and visual hallucination classically associated with an attack of temporal lobe epilepsy on the road to Damascus.) None of the twelve apostles wrote anything. They were like most people of their times: illiterate. The unknown authors called Matthew, Mark, Luke, and John were not eyewitnesses either, since these lost original texts bearing their names were written fifty to ninety years after the death of Jesus. The copies of these originals were not written by eyewitnesses but by others a hundred years after the events.

If you look at this odd collection of writings, one of the first things you will see is the lack of relationship between the Jewish Bible, or Old Testament, and the New Testament. Biblical scholars point out the dramatic differences in the wrathful, warlike God of the Old Testament, Yahweh, who made a covenant with one human clan and dictated the more than six hundred laws of Judaism in the Hebrew Bible, and Jesus the merciful, loving God of all peoples in the New Testament.

Jesus never mentions Yahweh in the New Testament accounts, even though he is actively reinterpreting these Jewish scriptures. The Jewish Bible never mentions Jesus by name. Why are these two testaments put together in one book? Who decided that they were part of one revelation?

The reason the Jewish Bible was included in the canon was the fact that Jesus was a Jewish rabbi, teacher, and interpreter of the Jewish scriptures. The Jewish authors of the New Testament were not writing history but theology to support their own version of Jesus and his teachings. Jesus' disciples were Jewish and saw his mission as the continuation of God's covenant to bring greatness and justice to the oppressed Jewish people. To promote his acceptance by the Jews, his followers reinterpreted language in the Jewish Bible in an attempt to "prove" Jesus was the Messiah promised by God. They continually tried to show how Jesus fulfilled the "prophecies" of the Jewish Bible.

Judaism was a well-established religion at the time of Jesus' birth and was respected and tolerated by the Romans. Heretical offshoots and marginal groups that were opposed or persecuted by the orthodox Jewish establishment were viewed as dangerous troublemakers by both the Jewish leadership and the Romans. It was important that Jesus be seen as part of this mainstream religion. The debate about whether Christianity was just a refined and updated version of Judaism or a new religion that included Gentiles began with Paul versus Peter and James. Ultimately, the Jews rejected Jesus as the Messiah, and Christianity developed an anti-Jewish point of view. In passing, it is interesting to note that American politicians are fond of referring to our "Judeo-Christian" heritage in an attempt to fuse very different religions and gain access to the contributions and votes of a politically active Jewish minority. From the landing of the *Mayflower* in 1620 until long after the ratification of the Constitution, Jews, Quakers, and Catholics were legally persecuted and killed. Yet some Christians insist on ignoring historical facts and rewriting history to establish the Christian religion as the central force behind the formation of this nation.

Who had authority to decide which of the many first-century texts are inspired and preserved in accurate form by divine action? Who had authority to decide to include or exclude a text in the New Testament and in the Jewish Bible? The canonization, the final selection of documents and writings to be included in the "official" or orthodox version of the book, took over one thousand years for the Jewish Bible, and the New Testament was not fully approved as the twenty-seven books Protestants now use until 367 CE.

Different Christian groups have different Holy Bibles. There is debate about why the Catholic clergy, charged with deciding which books would be part of the orthodox Bible, did not include some noncanonical books cited in the traditional Bible.[3] Protestants removed some texts from this Catholic canon in constructing their Bible. The order and number of books included vary. The Samarians have 5; Hebrews, 39; Protestants, 39 in the Old Testament and 27 in the New Testament; Catholics, 46 in the Old Testament and 27 in the New Testament; Greek Orthodox, 50 in the Old Testament and 27 in the New Testament; and Ethiopians, 81 books total. Even within the books selected there are multiple translations in English based on Hebrew, Latin, Greek, and/or previous English versions. Believers who want to use the Bible as evidence for any claim need to specify which Bible.[4]

The canonization of the Bible raises other potential problems for believers. What do they do if future excavations unearth well-authenticated texts that dramatically change or contradict existing versions of the Bible? There have been recent debates about the Gospel of Thomas, the Gospel of Judas, the Gospel of Mary, and the Book of the Stranger, and other Gnostic texts that are not part of the conventionally approved Bible.[5] Who were the authors of these texts? Are they the later transcriptions of oral histories of the apostles whose names they bear? Are any of the events or sayings in them historically accurate?

Irrespective of the Bible's imaginative and dramatic stories and its influence on the Western world, the Bible is a book like any other. It is a product of human authors and reflects their motives and experiences. Its authors were taken with Jesus' charismatic message, which each reported with his own spin.

If Jesus was literate, he never wrote anything. Now, if Jesus were God and his purpose in assuming human form was to "reveal the nature of God" to help his creatures comprehend his message, a reasonable person would assume he would write his message down himself in clear and unambiguous language. This is certainly within the abilities of an omnipotent creator. How do Christians explain why God instead selected over forty-eight human authors and divinely inspired them to write down over the course of 1,400 years (from about 1280 BCE to 200 CE) the most important message in creation? Why didn't God provide a clear and convincing seal of authentication, a public, undeniable well-documented miracle, for example, to let us unquestionably know his ultimate message? To do so would not interfere with our free will, because humans would be free to reject it.

According to the criteria of C. S. Lewis, endorsed by Collins, miracles are restricted to "great occasions, not of political or social history, but of that spiritual history which cannot be fully known by men."[6] The presentation and authentication of Jesus' message of salvation is the spiritual great occasion "par excellence," but no undeniable miracle is performed, extensively witnessed, and recorded that authenticates the texts as the word of God. Christians claim that that miracle is the Resurrection. But again, there are no records of the Resurrection except in the Bible itself. All we get is the circular reasoning that the Bible is God's message because "the Bible tells us so and as God's messenger it is infallible." This same claim is, of course, made by all sacred texts like the Qur'an and the Book of Mormon.

## THE RESURRECTION AS PROOF

The Bible is the source of people's belief in miracles, which supposedly demonstrates Jesus' divine nature. I discuss miracles in detail in chapter 3, so I will deal here only with the most essential Christian miracle, the Resurrection. Many books have been written on this topic, some of which I have read. In them, skeptics propose plausible, natural explanations for the details of the Resurrection story, as reported in the Bible. Almost all of these authors, critics and defenders, assume that most of the biblical text and descriptions are reliable. Defenders, mostly academic theologians, try to discredit these alternative explanations based on their own interpretations of the biblical texts and to argue that the best explanation for the Resurrection is a supernatural miracle. My analysis after sampling this literature is that this is a miracle story based partially on what probably happened and partially on fictional embellishments needed to create a miracle story. Jesus' followers constructed this legend to convince other people that Jesus was God incarnate. In fact, there is no evidence of this momentous event outside of the story in the Bible.

The first biblical text to claim Jesus rose from the dead is the Epistles of Paul. He gives no details of the events that were later described by Mark, Matthew, Luke, and John. Paul says in his Resurrection text, "For I delivered unto you first of all that which I also received, how that Christ died for our sins according to the scriptures" (1 Corinthians 15:3). This establishes the fact that Paul's belief is not based on eyewitness evidence but on second- and thirdhand reports. Some defenders claim, without any textual evidence, that Paul means he was provided information from Peter and James when he first visited them in Jerusalem three years after the event. That is still secondhand, even if true.

So Paul did not contribute evidence of his own experience of Jesus' appearances after death. It is almost certain that Paul had an attack of temporal lobe epilepsy on the road to Damascus and experienced visual and auditory hallucinations presumably sometime after

Jesus' death (Acts 9:3–9). He suddenly fell to the ground and was blinded by a bright light that was not seen by his companions. He had an auditory hallucination, hearing what he believed to be the voice of Jesus. There is no description of the appearance of Jesus, either as a ghostly spirit or as a flesh and blood human being. Paul was in a post-seizure state for three days before his sight returned. This was a private, subjective experience, and none of Paul's companions saw Jesus.

Paul also mentions that Jesus "was seen of above by five hundred of the brethren at once" (1 Corinthians 15:6). Again, it is probable that he did not witness this event but believed those who told him about it. There are no details about where or when this alleged event took place. Almost certainly there was no such event, or if it occurred it was another example of the well-established phenomenon of mass hallucination. Other than this one passage, there are no mentions of the supposed event in any Christian or secular documents. If the Gospel authors were aware of such an event, it is very unlikely they would not include it as both evidence of Jesus' divinity and a fulfillment of the Jewish scriptures.

In subsequent biblical versions of Resurrection events, there are mentions of various numbers of women, including Mary and Mary Magdalene, visiting Jesus' tomb. Appearances to some women and apostles are described. Paul makes no mention of any visitations or appearances to anybody at the tomb. Paul reports that Jesus appeared to Peter first and then to the twelve apostles. He learned this from the apostles or from believers who heard the apostles teach. Paul's account is only reliable to the extent his informants were reliable. Paul provides no independent evidence that his belief in the Resurrection was based on actual historical facts. If Paul's text is evidence at all, it is extremely weak evidence.

The next biblical description of the Resurrection appears in the final chapter of the Gospel of Mark. He uses just eight verses, numbers 1 to 8. A consensus of Bible scholars agree that verses 9 to 20 were a later addition to the original chapter by unknown scribes. Mark's account ends with three women—Mary, Mary Magdalene, and

Salome—finding the tomb open and empty. The only person they see is a young man (an angel?) who tells them to go tell the disciples that Jesus has risen. Mark concludes with the women leaving frightened and telling no one what occurred. For this reason, Mark's text provides no evidence for after-death appearances and ends with the empty tomb. The consensus of biblical scholars is that the appearances described in verses 9 to 14 were added by a different author to enhance the legend according to its later versions.

Matthew's Gospel, written a few years later, takes twenty verses to add additional details about the Resurrection, some of which are fictional embellishments to the legend. While in Mark's account the tomb was open when the women arrived, Matthew describes earthquakes and the descent of an angel who rolls back the stone. As Mary and Mary Magdalene are returning to tell the apostles that the tomb is empty, Jesus meets them. This is the first biblical report of a post-death appearance. Matthew also adds a detail about the tomb being guarded. This is highly unlikely. If, as Matthew claims, Pontius Pilate sought to forestall problems with his contentious Jewish subjects and actually assigned Roman guards, he would have expected them to ensure the body was not disturbed. Failure of the Roman soldiers to prevent the body from being stolen would have resulted in harsh public punishment. Pilate would also have ordered that the grave robbers be found and punished. No such events are described in the Bible or secular histories of the times.

Matthew's effort to depict the event as clearly miraculous requires that the Jewish elders make a request to Pilate for the Roman guard (Jews do not work on the Sabbath so the guards would have to be Romans or Gentile mercenaries) to prevent the disciples from stealing the body. If the body were stolen, the Jews feared some might believe that this was a miracle or sign of divine approval. If this was their fear, why does Matthew have the Jews say after the event that, in spite of the guard, the disciples stole the body? The presence of guards was to forestall one of the presumed natural explanations of the empty tomb. It is hard to believe that the Roman guards, the only firsthand witnesses of a man claiming to be God moving the stone and miraculously rising from the dead with awesome spectacular accompaniment, would not immediately become believers and spread the good news.

Matthew reports the second appearance of Jesus, which was to the eleven apostles in Galilee. Some of the apostles "saw him" and "they worshipped him," but some apparently did not see him and they "doubted" (Matthew 28:17) In contrast to the first bodily appearance, this seems to have been some kind of a ghostly apparition typical of mass hallucinations.

Luke, writing a few years later, covers the event in fifty-three verses (Luke 24). Luke, like Mark, has the stone rolled away when the women arrive and reports none of the miraculous prelude and opening by angels reported in Matthew's version. Unlike both Mark and Matthew, Luke reports no appearance to the women. Instead, Luke adds a new first appearance to two of Jesus' followers, Cleopas and an unnamed companion (Luke 24:13–31). This is a weird story of the followers' meeting with a stranger on their way to Emmaus (six miles from Jerusalem) and engaging in conversation with the stranger for several hours about their beloved leader, Jesus. It is not until they stop for the evening meal that they finally realize the disguised stranger is Jesus. The two return to Jerusalem and tell the eleven apostles that they have seen Jesus. Verse 34 is confusing since the straightforward meaning is that the two followers are speaking to the apostles, saying, "The Lord is risen indeed, and hath appeared to Simon." Simon was a common Jewish name and could have been the name of the second follower. However, Christian theologians interpret this as the apostles speaking to the followers and claiming that the apostles already know Jesus is risen because he has previously appeared to Peter. This supposed prior appearance is not reported by Luke so we don't know when or where it occurred.

As the followers report their experience to a group of the apostles, a skin and bones Jesus suddenly appears and to resolve their doubts, he eats some fish. Jesus walks with them to Bethany, a distance of about two miles, where he ascends to heaven. Incidentally is it curious that in verse 37 the apostles, on seeing Jesus, "were terrified and affrighted." Why, if they expected Jesus to return in three days as he prophesized, and they were his beloved companions, would they have anything to fear from the apparition? Didn't Peter, who supposedly experienced the first appearance, tell the rest of the disciples Jesus was risen after he had seen him?

Finally, we have the most elaborate account in the fifty-six verses of John's chapters 20 and 21. John's account describes Mary Magdalene alone visiting an already open tomb. There are no angels mentioned. She is not informed that Jesus has risen and wonders what the authorities have done with the body. Peter and "the other disciple, whom Jesus loved" visit the tomb. After they leave, two angels appear and Mary turns and sees her beloved Jesus, but she assumes the man to be a gardener. When Jesus speaks, she recognizes him and reports the appearance to the apostles. Shortly thereafter Jesus appears to ten apostles. Eight days later he reappears to "the disciples," including the doubting Thomas. To add additional support to his case, John includes an appearance of Jesus to about six of the disciples while they were fishing. After a miraculous catch, Jesus eats fish and bread with the whole group.

The Acts of the Apostles, traditionally ascribed to Luke, mentions that Jesus' appearances occurred during the forty-day period after his death (Acts 1:3). None of these appearances are described in detail. Who was present and what did the risen Christ say or do? This would not be independent evidence of a resurrection. None of these later appearances are described in the Gospels. If they occurred, why not? Even better evidence that the Resurrection is a legend is the fact that neither of the earliest versions, Mark and Matthew, mention the most convincing miraculous appearance—to the apostles in a locked room—described in Luke and John.

Let me repeat, the New Testament was not written by eyewitnesses recounting their experiences. It was not written as a biography, a history, or a scientific text. It was written by believing Christians to document and promote their own versions of Jesus' teaching at a time when there were many different Christian theologies in competition. While the writings are based on a core of actual historical events, details about these events are unreliable and fictitious elements have been added. All descriptions of historical events are only probable approximations of what actually occurred.

Let us examine which events, following Jesus' death, have a high historical probability of being true and are needed to produce a reasonably reliable and complete account. First there is a reasonable probability that someone representing the Jewish authorities buried Jesus in a tomb and that one or more of Jesus' female followers found it empty. It is also probably true that one or more of Jesus' disciples had an experience that they interpreted as a postdeath appearance of Jesus.

How are these events to be explained using only natural phenomena and without postulating they were the result of a highly improbable supernatural miracle? Jesus' burial in a tomb was done in conformity with the sensibilities of the Jewish elders, the Sanhedrin, about the Passover and to prevent his followers from stealing the body either to claim resurrection or to give it an honorable burial. The Jewish agents most probably unceremoniously deposited the naked body in a convenient empty tomb owned by the Sanhedrin member Joseph of Arimathea. It is inconceivable that Pilate would have consented to the honorable burial of someone threatening Roman power in rebellious Palestine. The body was therefore probably undisturbed on the Saturday Passover from sunrise to sunset.

After sunset, it is likely the Roman or Jewish authorities rolled back the stone, removed the body, and deposited it in an unmarked grave where criminals were buried. The women reported the empty tomb to the apostles, who were hiding in fear of the Jews. One of the grief-stricken women, perhaps Mary Magdalene, could have had a hallucinatory experience interpreted as an appearance, or this detail could have been added to the story as it was developing. It is probable that the apostles had at least one, or less likely a number of hallucinatory experiences interpreted as the appearance of their beloved leader.

According to the biblical account, they remained in hiding until fifty days after the Crucifixion. Then, on Pentecost, the "Holy Ghost" inspired them to go out and spread Jesus' message and the good news of his Resurrection to a dispirited group of followers (Acts 1:3). The period of hiding is reasonable and likely under the circumstances. Based on this tiny core of probably true events, this natural account explains all the accepted facts. There was no Resurrection and it is not necessary to invoke a miracle to explain it. Using this core of likely events, the legend began to grow, gaining fictitious details to make it more dramatic and believable to prospective converts. There were probably no guards at the tomb and the number of separate appearances was increased in the telling and retelling of the story. Over thirty years, all these embellishments were combined to create the legend that reached the ears of the Gospel writers who used them to validate their various theological perspectives.

This is, however, only one plausible natural explanation of an unbelievable, biased text. In truth, all the reconstructions of believers and skeptics are speculations based on subjective interpretations of probability. I will not, therefore, include a detailed defense of my scenario but will only respond to some frequently stated defenses of the miracle.

Theologians argue against the existence of mass hallucination, but objective reports document such occurrences. There may have been private hallucinations, particularly during Jesus' immediate postmortem period. Even today, there are reports of several individuals seeing flying saucers or Bigfoot or Yeti or UFO crashes in Roswell. Which is more probable: mass hallucinations or someone rising from the dead? Investigation of such occurrences has documented the belief in the hallucinatory exaggeration of the physical event. Such events have the ability to generate a powerful emotional experience that persists after the experience has ended and will be retained in spite of clear and compelling evidence that the physical event never occurred. A detailed defense of the hallucinatory explanation can be found in *The Empty Tomb: Jesus beyond the Grave*.[7] There is no reason to believe all the descriptions of multiple appearances since it is the nature of legend to expand on the initial core. The appearance to James, the brother of Jesus, is reported secondhand in Paul, not documented in any other Gospel, and could be the common experience of a bereaved relative.

But theologians ask, why didn't the Jews simply discredit the apostles' claim by producing the body of Jesus? Because the apostles waited fifty days before they began to publicly make this claim. Any body would have decomposed beyond recognition. We have no way of knowing when or if this claim was called to the attention of any of the Jews or Romans who had knowledge of where the body was deposited. There is no history or tradition of Christian veneration of the tomb site by the apostles or early Christians.

The theologians persist: "If the Resurrection was a hoax or never occurred, why would the followers of Jesus give up their lives for a lie?" But, did they willingly give up their lives rather than denounce the Resurrection? Do we know exactly what beliefs or practices they were asked to denounce to be spared? How do we know if they were given the opportunity to save themselves at all? Do we have independent sources other than Christian tradition? Obviously Jesus' followers did not think it was a lie. They believed, without good evidence, that it occurred. They did not fear death because they believed that a martyr's death at the hands of their Jewish and Roman persecutors was the key to the everlasting kingdom of bliss. The strength of a belief is not necessarily related to its truth. Like the apostles, Islamic suicide bombers and Japanese Kamikaze pilots have died for causes in which they strongly believed. If the apostles saw something after Jesus' death, it was not a living Jesus but a ghostly hallucination. Few people would die to preserve what they know to be a lie, but a firmly held false belief is no lie. So, the death of Christian martyrs proves only their faith in the Resurrection, not that the fact of a Resurrection was real.

The story of the Resurrection was not necessarily created as a deliberate hoax. The authors of this story were building a legend to reinforce their interpretation of the teachings of their much-admired leader, Jesus. How could they continue to spread his message after his public execution and disgrace? We see a natural progression of their th "His spirit must live on"—"Even his ghost continues to teach us"—"His ghost assumes bodily form to console us"—"In fact, we have seen his resurrected body!" Each time the story was retold, more "evidence" was included to prove it was a real miracle. The grave is sealed, the guards are added. The Jews invent a lie to explain it. The numbers of visitors to the empty tomb is increased, the contents of the tomb are described, and more appearances to more people are added.

All agree that the stories told about Jesus were transmitted by word of mouth from the time of his death about 30 CE to at least 60 CE, the earliest date for the writing of the Gospel of Mark. We've all played the children's game in which a message is sent, via a whisper, through a chain of people. To everyone's amusement, the message is always distorted by the time it reaches the last person. Imagine the distortions that would occur when trying to transmit a complex message over decades! Repeated scientific investigation and common experience with the verbal transmission of factual information has always confirmed that it does not take many repetitions to accumulate significant errors—both leaving out important details and adding new imaginary information. Those who argue that thirty years is too short a time for this legend-building process to occur are just wrong and ignore scientific evidence and common experience. Even with the special care given religious stories and even with the use of memory aides, it is highly unlikely that events detailed in the Gospels, including the Resurrection, were complete or accurate.

## COLLINS'S DEFENSE OF THE BIBLE

Suppose for a moment that Jesus actually existed, spent his short life in Palestine, taught a version of Judaism that stressed repentance for sin and devotion to Yahweh, and that he was crucified, but no one bothered to record the events of his life and his teaching. Suppose no stories were collected and preserved. Suppose there was no Bible.

How would you know that Jesus was the one true God? Would Collins have any reason other than the biblical accounts to seek fellowship with Jesus? I don't think so.

This is what makes the accuracy of the Bible so essential to the truth and believability of Christianity. The whole game is riding on it. Christians are literally betting their lives that the Bible is historically true and without error. Once you question the literal truth of the Bible—say, for example, you believe that casting out demons represented some psychological treatment of mental illness—then how can you know with certainty that Jesus' appearances after death even occurred or were not hallucinations? How do you know whether Jesus claimed to be one of the human sons of God or God himself in human form? If the Bible has one error, isn't it likely to have more?

If you want to consider the Bible as more than just a human book and as the most important set of divinely inspired documents ever written, then you should have much better evidence of its accuracy than exists for other, less important ancient documents. Any analysis of the Bible's age, authorship, and historical context should be convincing and correct. However, even meeting these conditions does not establish the truth or accuracy of the Bible's contents. The details reported would require independent support by appropriate, quantitatively and qualitatively sufficient secular evidence, such as less biased textual resources, to validate their accuracy. The believers and religiously inspired clergy, who claim the authority of the Bible as evidence of their claims, misrepresent and misinterpret the large amount of scientific and scholarly evidence about who wrote it and why they wrote it, and who translated it and why they wanted a translation, and who decided what texts should be included and why they should be included.

Collins mistakenly states, "Concerns about errors creeping in by successive coping or bad translation have been mostly laid to rest by discovery of very early manuscripts."[8] This unsupported assertion dismisses years of detailed studies by Bible scholars who have documented such errors by reading and comparing texts in their original languages.

The discovery of the Dead Sea Scrolls, which were older than the previous texts of the Jewish Bible; the manuscripts in the Nag Hammadi Library, a collection of forty ancient texts found in 1947; and other early manuscripts raise serious questions about the accuracy and motivations of those who copied and translated the version of the Bible we read today. The Dead Sea Scrolls were written from 250 BCE to 70 CE by a Jewish sect, the Essenes, who lived in the Judean desert. This period spans the lifetime of Jesus, yet he is never mentioned in any of the hundreds of writings found. Biblical studies continue to provide compelling reasons to question the historical accuracy of today's Bible. Critics of evolution who point to "gaps in the fossil record" ignore the gaps in our knowledge of the history of Judaism and Christianity and their respective Bibles.

To prove his point, Collins cites only the standard text *The New Testament Documents: Are They Reliable?* by F. F. Bruce, which attempts to show when the documents were written and who composed them.[9] Bruce admits there is no way to prove the historical truth of the texts in detail, especially the miracles they describe. If one already accepts the divinity of Jesus, then, Bruce claims, miracles are in keeping with his divine message. But this is a question-begging response. If the question asked is whether the Bible can be used to prove Jesus was divine, one cannot then use the assumption of Jesus' divinity to authenticate the Bible. Bruce regards the Bible's authors as completely trustworthy, in the sense that they were not deliberately lying. But that does not mean they did not embellish the stories they were told by the supposed eyewitnesses, nor does it prove that the eyewitnesses accurately and completely reported the events they interpreted as miracles.

Collins also includes in his bibliographical notes *The Historical Reliability of the Gospels* by Craig Blomberg and *The Historical Jesus* by Gary R. Habermas.[10] While these works do a good job of establishing certain core facts, such as that Jesus was a real person who lived, taught, and died on the cross in first-century Palestine, they again fail to satisfactorily resolve textual contradictions or validate the supernatural birth and Resurrection of Jesus. Blomberg says, "Christians may not be able to prove beyond a shadow of doubt that the gospels are historically accurate, but they must attempt to show that there is a strong likelihood of their historicity. Thus the approach of this study is always to argue in terms of probability rather than certainty, since this is the nature of historical hypotheses, including those which are accepted without question."[11]

In discussing differences between the synoptic Gospels, Blomberg admits, "Many of the differences among the parallels are insignificant; the gospels do not claim to supply a highly literal translation of Jesus' words. Besides being translated into Greek, Jesus' teachings have been freely paraphrased throughout the gospels as a glance at any synopsis reveals."[12] The contradictory passages in the synoptics are not explained but are rather explained away as "apparent" contradictions. Blomberg concedes, "The solutions here may not carry equal conviction; scholars have often put forward different reconciliations for many of them. In a few cases, better solutions may await future research."[13]

When dealing with the differences between the synoptics and John's Gospel, he glosses over real historical and theological differences by developing more or less plausible solutions. These are based on his preconceptions that the truth of the Bible can stand up to critical analysis. He succeeds only in establishing the strong probability that Jesus lived in Palestine from about 3 CE to 30 CE; was a rabbi, a teacher of Jewish scriptures and religion; and was crucified. The general outlines of his teaching and Jesus' travels are available in the text. The authors of the text reported Jesus performed miracles, but there is no evidence from any independent source that any miracles were actually performed. In short, the Bible has a historical core to which was added different theologies and elements of legend.

Habermas states his bias in his introduction: "This writer does not wish to be a part of such efforts that teach or even imply that Scripture is not a sufficient basis for Christian belief. This book is devoted to developing a new area of apologetics and not to questioning the basis of Scripture. In fact, this writer believes that the best approach to apologetics (in general) is one that begins with the evidence for the trustworthiness of Scripture and then proceeds on this basis."[14]

Habermas is concerned with presenting evidence for the historical existence of Jesus, and, in this respect, his evidence is convincing. He is less convincing when recounting the events in Jesus' life, specifically the virgin birth, the miracles, and the Resurrection. There is no doubt such events were reported to have occurred by the authors of the New Testament, but there is little credible evidence that their interpretation and detailed descriptions are correct. He correctly points out, "History cannot reach a point where it is positive of its findings in all instances. As with physics, medicine and other indicative disciplines there is a certain amount of dependence on probability in history as well."[15] Habermas assumes that if the biblical authors reported a miracle, it is probably a historical fact, discounting the improbability of miracles occurring at all. His burden of proof is too easily met.

Bruce and other defenders of the reliability of the New Testament routinely make much of the claim that, if the authors were not eyewitnesses, there were many eyewitnesses of the events still alive at the time the Gospels were first written. This claim does not provide as much support for the detailed historical accuracy of the texts as they might hope. Jesus' disciples were likely in the same age range as Jesus, say, from twenty-five to thirty-five, when he was crucified. The average life span for a first-century Palestinian has been estimated from thirty to forty-two years, so you would expect most of them would have died within thirty years of the Crucifixion. The same could be said for most of the adults who were eyewitnesses of some portion of Jesus' life.

Furthermore, the apostles were wandering beggars subject to the physical demands of foot travel and malnutrition and therefore probably had shorter than average lives. They also apparently faced premature death due to persecution. James was reportedly martyred in 61 CE. So when the first of the Jesus stories, the Gospel of Mark, was written in 60 CE, the few living eyewitnesses would have been forty-six to ninety-five years old. Finally, Jesus never left Palestine, so the eyewitnesses would have been concentrated there. Few would have traveled to the cities of Asia Minor where Paul was teaching his version of Christianity or to Rome where Mark was thought to have written the first Gospel or to Syria where the book of Matthew was believed to have been written.

The authors of the Gospel stories that described Jesus' miracles and message were devout followers of Jesus. Each had his own point of view and wrote to influence a special audience of believers. The Gospel authors were clearly biased theologians and not objective historians. How reliable a picture of my life, philosophy, and actions would you get if the only record of them was a series of books written thirty years after my death by people who love and admire me?

Most pious Christian scholars conclude that the texts selected for the New Testament have come down to us substantially but not exactly as they were first written and are generally but not universally in accord with some known facts about historical people and places. This kind of general, factual core does not mean that the virgin birth or Jesus' sayings or miracles or his Resurrection are true historical events. The New Testament is not a work of fiction but it is also not a history, although it contains historical references.

Collins cites Lee Strobel, a lawyer who is not an expert in Bible scholarship, history, or archeology, as the defender of the historical accuracy of the Bible. As an attorney, Strobel cherry picks his evidence. His short booklet consists in interviews of and selected quotations from Christian scholars who all assume that the details in the Bible are historically accurate.[16]

No Jewish contemporary of Jesus, such as the writer and historian Philo-Judaeus, or any Roman historian, such as Tacitus, mentions Jesus, his miracles, or his Resurrection. Why not? The earliest reference is by Flavius Josephus (33–100 CE), a Jewish general who defected to the Romans and wrote the secular histories *The Jewish Wars* and *The Jewish Antiquities*. In the latter there is only one paragraph that mentions Jesus, which most scholars believe to be at least in part a Christian forgery. While this paragraph might be used to support the existence of Jesus and his Crucifixion, it cannot confirm his miracles or his Resurrection. Habermas concedes, "However, we still cannot conclude that ancient extrabiblical sources, by themselves, historically demonstrate the resurrection, as is true for Jesus' crucifixion."[17]

Later second-century secular references to Jesus are based on earlier unverified accounts by Christians. There are no other secular writings that independently authenticate anything about Jesus. From these combined resources, all we can conclude about the historical Jesus is that, inspired by John the Baptist, for a short time he preached a message of repentance for sin as preparation for an imminent apocalyptic return of the God of the Jewish scripture, and that the Romans crucified him as a troublemaker.[18] Given the earthshaking importance of Jesus' mission, this lack of secular interest strikes me as another reason to be skeptical of biased and unreliable biblical accounts.

Defenders of the Bible claim that the evidence of its historical accuracy is better than that for other secular histories and documents from ancient Rome and the Middle East. They argue that these secular manuscripts are not available in original texts and that there are fewer copies available than the Bible. They also note that the copies of secular documents that we have were written longer after the originals than is the case for the Bible. This may be true, but Caesar's *Gallic War* and the histories of Herodotus do not make the extraordinary claims of the Bible and do not aspire to its everlasting relevance and importance. Julius Caesar did not perform miracles or rise from the dead. Scholars do not accept these secular manuscripts as accurate in every detail. As Habermas correctly observes, "There is always a subjective element in reporting the past and conclusions from this discipline must be couched in probabilistic terms."[19] Little is riding on whether the details of secular histories are accurate, but a whole worldview is dependent on the accuracy of the Bible, which must therefore be examined using a higher standard of evidence.

No convincing scientific evidence exists to verify the biblical stories of the Jews' Egyptian captivity, the forty years of wandering in the desert, the Flood and Noah's Ark, or the real existence of Abraham, Job, and Moses. Christians tend to treat the Jewish Bible allegorically. The allegorical interpretation of parts of scripture allows defenders to save the parts they want to use historically but raises the question of how you can reliably differentiate the historical from the allegorical parts. How do we distinguish a correct interpretation of an allegory from a misinterpretation? At various times, Christians have used biblical interpretations to support almost anything. Persecution of Jews as Christ killers was justified by the anti-Semitic writing of Matthew, for example, 27:25; execution of witches by Exodus 22:18; execution of homosexuals by Leviticus 20:13; male domination of women by Genesis 3:16; slavery by Genesis 9:25, and so on. Today, we know this was wrong, but during the course of history many Christians believed that these discredited beliefs were the "true" interpretation.

Most Christians practice "cafeteria Christianity," selecting parts of the Bible that are acceptable to them and ignoring parts they find unacceptable. They readily cite verses that clearly urge loving, compassionate, moral behavior. However, no moral, compassionate human being would want to live in a world where every biblical command was followed rigorously. While Christians include the Jewish Bible in most of their versions of the Bible and accept the Ten Commandments, most choose to ignore other biblical commandments as the inspired word of God. Biblical accounts tell us Jesus urged his followers to observe the Mosaic Law, and in Matthew 15:17–20 we read:

> Think not that I come to destroy the law, or the prophets: I am come not to destroy, but to fulfill. For verily I say unto you till heaven and earth pass, one jot or one tittle shall no wise pass from the law, till all be fulfilled. Whosoever therefore shall break one of the least commandments, and shall teach men so, he shall be called the least in the kingdom of heaven: but whoever shall do and teach them, the same shall be called great in the kingdom of heaven. For I say unto you, that except your righteousness shall exceed the righteousness of the scribes and Pharisees, ye shall in no case enter into the kingdom of heaven.

The Bible tells parents how to deal with a rebellious and stubborn son—stone him to death (Deuteronomy 21:18–21). Furthermore, adultery (Deuteronomy 22:22 and Leviticus 20:10), rape (Deuteronomy 22:25), a wife who lies about her virginity or false accusation of sex before marriage (Deuteronomy 22:13–22), cursing your father or mother (Leviticus 20:9), incest (Leviticus 20:11), homosexuality (Leviticus 20:13), and working on the Sabbath (Exodus 31:14–17) all merit the death penalty. Are these parts of the Bible just wrong? Were they right before Jesus was born and wrong now?

As a moral guide, the Bible gives mixed messages. Collins does not discuss whether the Jewish Bible's Mosaic Laws apply to Christians. But this has been a perplexing, unresolved issue for Christians since the so-called Council of Jerusalem, which is alleged to have occurred between 50 and 62 CE. This council was convened to discuss disputes between Paul and the more traditional Jewish apostles. Paul, as apostle to the Gentiles, argued that the Gentile converts did not have to be circumcised or observe Jewish dietary laws. But the more traditional Jewish apostles held that the laws of God and the Jewish scriptures are eternal and apply to all Jews and new converts for all time. A temporary compromise was reached that allowed Gentiles to be admitted to the church if they observed some of the traditional Jewish laws. Even so, the Jews ultimately disowned the Christians as a heretical cult. The Bible supports both Paul (Romans 8:4–6) and the Jews (Matthew 5:18–19), leaving the question of the observance of Jewish law unresolved.

The New Testament is also contradictory when it comes to moral guidance. The fifth commandment is "Honor thy father and thy mother." But Jesus is reported as saying, "If any man come to me and hate not his father, and his mother, and wife, and children, and brethren, and sisters, yea and his own life also he cannot be my disciple." This is one example of the many contradictions and historical errors in both the Jewish Bible and the New Testament.[20] Do these irresolvable discrepancies make us more or less confident of their divine origin?

Another often-ignored aspect of the Bible is the frequent description of sexual activity, some approved and some disapproved by the authors.[21] Why is God so interested in our sex drives and reproductive behavior? What does the Bible really say about contraception, masturbation, artificial insemination, polygamy, divorce, prostitution, and abortion? I wonder why Yahweh attached such importance to circumcision? What cosmic divine purpose was uniquely served by cutting off the foreskin? Why was this sexual location selected for ritual identification of God's chosen people, instead of, say, shaved heads, a ring through the nose, or scarring the abdomen?

There are lots of first- and second-century writings about Jesus that were excluded by the orthodox Catholic authorities who decided on what would be in the canonical Bible. One piece of evidence that seems to convince Collins of the accuracy of the New Testament is that we have many first- and second-century copies that tell the same stories. But that is the purpose of copies. There is a great deal of variation that occurred in copying, but there are also a lot of passages that match word for word. However, if the initial version is wrong or inaccurate in any detail, copies will only reproduce the errors. Even identical copies of, for example, Mark dating back to 30 CE would not establish what was really true about the New Testament stories.

As a scientist, I would expect Collins to be more skeptical and demand some really good evidence before he accepted the New Testament stories about Jesus as true. I have read the Bible and many books about the Bible. Objectively, there are only a few things we can assume with reasonable confidence from this book: Jesus was crucified in about 33 CE. People who considered themselves his followers told stories about his life and teachings for twenty to thirty years after his death. Some stories were told by people who had seen him; others were second- and thirdhand accounts. Few people could read or write, but some of these stories about Jesus began to be written down about 60–75 CE by an unknown author who was given the name of the apostle Mark to give his version more authority, as was the custom at the time. This author wrote down stories he heard about a teacher, a rabbi, Jesus, whom he believed was a good and wise man, the Messiah sent from Yahweh to save the Jewish people. "Mark" wanted to convince his readers that his version of Jesus' life and mission was true. Two authors who used the names of the apostles Matthew and Luke read and partially rewrote Mark's story between 90–100 CE. They were also believers in Jesus' teachings.

New Testament defenders pose another question. If the original texts contained errors, why didn't the people who witnessed Jesus' activities and teaching write texts to correct the errors? I don't think this is a strong reason to accept the texts as the word of God. We do not know if these early Gospels were widely circulated and read, but we know that most eyewitnesses in first-century Palestine were illiterate, and few spoke (and fewer wrote) Greek. We do know that there were other things on the minds of Jews in the period 66–70 CE when the New Testament was first written, because the country was in revolt against the Romans. So, it is not surprising that no commentary on these texts was written until the first and second century.

To give you an example of the dubious nature of eyewitness information, there are books about alien abduction that include numerous eyewitness accounts. They can be shown to have been written by specific authors who lived at the same time as the eyewitnesses whom they interviewed. Dates and places can be verified. Comparisons can be made between different accounts of specific abductions, sightings, and so on. Reports about and belief in alien abductions are both historical fact. What is not a proven historical fact is that whatever occurred to initiate these reports was an actual alien abduction. In fact, based on all objective investigations and known science, there is little reason to accept that any alien abductions have ever occurred.

Believers may admit that the New Testament stories were probably not eyewitness accounts but they assert that there were eyewitnesses alive who could have challenged these accounts when they were circulated. Disputing these New Testament writings after their appearance is easier said than done. There were only a very limited number of copies available for many years. Initially these New Testament stories were only verbally spread in Jewish synagogues and homes with limited audiences. Travel in the first century was difficult, and it took time for these texts to reach the areas where the eyewitnesses lived. We have no way of knowing how many of these eyewitnesses of biblical events were alive or were aware of the existence of the original writings. The majority of Jews rejected the apostles' message, so they challenged accounts by Jewish eyewitnesses. Given the difficulty of preservation and discovery of documents, the absence of written challenges to biblical accounts does not mean there were no eyewitnesses with different narratives. We do know that in the immediate aftermath of Jesus' death, many versions of his story and teachings competed.[22]

These problems have eroded belief that the Bible is without error. In a 2006 Gallup Poll, only 28 percent of Americans expressed belief in the absolute truth of the Bible, 49 percent believed it was divinely inspired but not literally true, and 19 percent agreed with my position that it is a book of fables, legends, and history.[23] In a 1993 international comparison, the only countries with a higher percentage of belief in biblical inerrancy were the Philippines (53.7 percent) and Poland (37.4). Less than 10 percent of people in Great Britain, the Netherlands, Russia, and Germany believed it.[24] Subsequent polls have documented a gradual trend of increasing doubt about the absolute truth of the Bible in the United States. In 1980, 40 percent believed the Bible was the actual word of God, and 10 percent thought its stories were fables. By 2007, only 30 percent accepted it as God's word, and 22 percent thought it was just a book.[25] When broken down by education, 60 percent of those without high school education, 42 percent of high school graduates, and only 20 percent of college graduates believed the Bible was the actual word of God.[26]

## THE BIBLE AND SCIENCE

When both personal experiences and the rational arguments of science contradict or are incompatible with the Bible, you have only four ways to resolve the problem:

- Science has made errors in observation or interpretation of nature because the Bible contains no errors or contradictions.
- Believers have made errors interpreting the Bible, and there is always a new correct interpretation that can be made that preserves the Bible free of errors and contradictions.
- The apparent contradictions are due to the reader's inability to consult the true, original text of the Bible, which is lost to us.
- The Bible is a human book, contains errors and contradictions, and is not the standard for truth about things as they exist in the real world. Accept the science as a better approximation of the truth about the real world.

Collins agrees that some parts of the Bible cannot and should not be treated as accurate history or science, and states, "Texts that seem to describe historical events should be interpreted as allegory only if strong evidence requires it."[27] Apparently, strong evidence exists, in Collins's view, when broadly accepted scientific explanations clearly contradict or are incompatible with a literal interpretation of the biblical text.

Collins rightly notes that the Bible was not written as a science textbook. The Jews for whom the Old Testament was written would not have understood big bang cosmology or the evolutionary development of life-forms. But if science and the divinely inspired text of the Jewish Bible are both ultimately true, shouldn't there be a better correspondence between them on the facts of science? Collins admits, for example, the two creation myths in Genesis (Genesis 1 and 2) are contradictory in details. While I agree the purpose of the authors of the Jewish Bible was to impress their readers with the awesome power of Yahweh and not to create understanding of the scientific laws of nature, their writings contain serious errors that a divinely inspired writer would not make. As a geneticist, Collins can undoubtedly see the authors' error in Genesis 30:31–43, when they described how exposing livestock to different colored rods influenced the color of the livestock's offspring.

As for the New Testament, the science is no better. Why do the authors describe Jesus as casting out demons? Science does not attribute mental illness to demons, and doctors do not practice exorcism. If the Bible is divinely inspired, why would Jesus permit the perpetuation of this error? I know Jesus' purpose was religious instruction and not to provide scientific information, but why do the biblical authors have Jesus continue to mislead his audience about demons when it would have been just as easy to conform his teaching to the true facts of nature? The answer that seems most plausible to me is that the authors of these legendary stories about Jesus were limited in their knowledge of science. God did not prevent the errors because God had nothing to do with the production of the text.

Many Christians have attempted to use science to validate the historical accuracy of the sacred texts. However, this is a double-edged sword. If, for example, the parting of the Red Sea to save Moses or the Flood of Noah were to be explained by natural causes, then it removes the biblical support of miraculous supernatural causes. This results in a kind of reversing of the God of the gaps. Such "Christian science" moves closer to belief in a deist god, an impersonal creator who sets natural processes and laws in motion and then lets the universe evolve without further interaction.

While using archeology can establish that a few of the places and people mentioned in the Bible actually existed, it cannot reproduce what was said and done in detail. While some archeological data are consistent with some biblical passages, other archeological data shows the biblical version of history to be erroneous. As a result, science-savvy Christians, who hope to find scientific support for the Bible, are faced with the task of continuously reinterpreting texts to make them consistent with the latest scientific discoveries.

Collins is correct in pointing out the dangers of this quest for scientific support but defends his faith by claiming that the supernatural is beyond the scope of science. "The meaning of human existence, the reality of God, the possibility of an afterlife, and many other spiritual questions lie outside the reach of science."[28] This is a separation of scientific knowledge and religious faith, not a demonstration of their compatibility. He does, however, attempt to avoid inconsistencies with scripture and evolution by reinterpreting it himself. Collins sees that once you start down this slippery slope, you can "eviscerate the real truths of faith" and will need to decide "where to place a sensible stopping point."[29] Unfortunately, Collins's stopping point is the use of flawed and unreliable written reconstructions of alleged eyewitnesses to the life of Jesus.

## THE PROBLEM OF GENESIS

Collins is concerned with the 49 percent of American Christians who find Genesis a barrier to accepting the theory of evolution. He borrows an allegorical interpretation of Genesis from St. Augustine. Realizing the inconsistencies in his own interpretation, Collins admits, "Despite twenty-five centuries of debate, it is fact to say that no human knows what the meaning of Genesis 1 and 2 was precisely intended to be."[30] However, when he says, "We have still not resolved the public controversy about evolution nearly 150 years after Darwin's publication of *The Origin of Species*,"[31] he needs to clarify that this is a public-relations controversy, not a scientific controversy. The essentials of evolution are accepted by all but a marginal few in the scientific community. The only scientific controversy is in expanding and applying the theory and filling out the details of evolutionary mechanisms. No scientific explanation is ever complete and unchallenged.

It has been previously noted that, as a scientist, Collins sees the difficulty of a literal interpretation of the Bible, especially Genesis, and its two contradictory creation stories. While he is willing to concede that this is poetic literature, he does not discuss why God fails to give us a literal and accurate account of his creation. As Collins correctly observes, God, or his surrogates, was not writing a science text for first-century humans; still, couldn't God have at least gotten the time scale right without reference to "radioactive decay, geologic strata and DNA"?[32]

Once you allow for allegorical interpretation, factual error, or contradiction in the Bible, the whole content is suspect and cannot be used as strong or appropriate evidence in the debate with unbelievers. If Collins wants to accept the Bible as divinely inspired scientific evidence of a personal God, he needs to give us the criteria by which to determine which parts are objectively, historically accurate and describe real events and actual quotations, and which parts are legends, myths, morality plays, allegories, and metaphor. Which are objective truths and which are literary devices that can be subject to multiple subjective interpretations?

It is very easy to explain Genesis and Adam and Eve. The story is an ancient creation myth of the Middle East committed to writing by a very creative ancient writer without benefit of modern science. As Collins attempts to reconcile an allegorical interpretation of Genesis with the scientific truth of evolution, he encounters the twin problems faced by anyone who tries to fit the Bible to the facts of history and science: inconsistency and incoherence.

The explanation that Collins and his mentor, C. S. Lewis, suggest is that Adam and Eve were not the only occupants of the Garden of Eden. This allows for Cain to have a wife without incest but creates havoc with the doctrine of original sin. Central to almost all Christian sects is the idea that Adam and Eve were the only humans created by God and that they brought sin and evil into the world by violating a specific divine command. God threw them out of the garden and condemned them to painful childbirth, a life of hard work, and loss of immortality.

The Hebrew word for Adam is the noun "humankind," and Adam, therefore, was interpreted as representing the human race.

According to the Bible, the penalty for this original sin falls on all Adam's "seed," that is, all humans. This, we are told, is why we need redemption and salvation to wash away the stain of original sin. If the sin had been Adam's alone, why would a just God impose the penalties on his innocent progeny? Why are we not born sinless and earn or lose our heavenly reward based on our own behavior? Christians cannot answer these questions without giving up the dogma of original sin and with it the need for Christ the redeemer.

Collins and Lewis propose the Adam and Eve story is an allegory for a special act of creation that bestowed higher-level consciousness on several hominoid animals (*Homo erectus*? *Homo habilis*?) who lived in the Garden of Eden. This group was tempted by some unnamed "someone or something" to become gods (*I guess they have problems with talking snakes*). Lewis's imagination (he created mythological worlds such as Perelandra, an ocean world with undulating pleasure islands) is on full display in his science fiction essay titled "The Fall of Man." In it he proposes a primitive form of man for whom "this new consciousness ruled and illuminated the whole organism, flooding every part of it with light and was not, like ours, limited to a selection of the movements going on in one part of the organism, namely, the brain. Man was then all consciousness."[33] Lewis proposes that God first created a supernatural race of Adams whose primordial consciousness was flooded with light. (*"Flooding the organism with light" does not seem to be a very illuminating statement. It explains nothing and is an unintelligible abstraction. It is not the light familiar to physics.*) According to Lewis, this primate with consciousness was a separate "species" with extraordinary power. He says, "His organic processes obeyed the law of his own will, not the law of nature."[34] "The length of his life was largely at his own discretion."[35] (*This would explain how Adam lived 930 years* [Genesis 5:3].) He had "a mysterious power of taming beasts,"[36] he existed in repose, in the joyful contemplation of obedient love and self-surrender. While "he had therefore no temptation in our sense to choose the self,"[37] only to continuously repose in God, "someone or something whispered they could become as gods."[38]

Lewis admits, "We have no idea in what particular act or series of acts the self-contradictory, impossible wish found expression. For all I can see, it might have concerned the literal eating of a fruit, but the question is of no consequence."[39] As to why the sin of pride or disobedience of these primordial humans, or Adams, should continue to exact a punishment on us today, Lewis does not feel it is a punishment but opines, "It may be that the acts and sufferings of great archetypal individuals such as Adam and Christ are ours, not by legal fiction, metaphor, or causality, but in some deeper fashion."[40] (*What deeper fashion?*) "I have thought it right to allow this one glance at what is for me an impenetrable curtain."[41] With this verbosity, Lewis runs out of ideas and words, stating, "With this I have said all that can be said on the level at which alone I feel able to treat the subject of the Fall."[42]

Lewis continues avowing that he has "the deepest respect even for Pagan myths, still more for the myths in Holy Scripture." He continues, "I do not doubt that the version which emphasizes the magic apple and brings together the trees of life and knowledge contains a deeper and subtler truth than the version which makes the apple simply and solely a pledge of obedience. But I assume that the Holy Spirit would not have allowed the latter to grow up in the Church and win the assent of great doctors unless it was also true. And useful as far as it went."[43] It is exactly the existence of "the Holy Spirit" that is in question. Adam commits the first sin, which was "very heinous." But, as Lewis says, "This is, if you like, the 'weak spot' in the very nature of creation, the risk which god apparently thinks worth taking."[44]

If the infallible God that Collins and Lewis worship exists, this God does not take risks. God never loses a gamble. Humans take risks when they bet on an unknown and uncontrollable future outcome. The crooked roulette operator does not risk anything. God was certain before Adam was created and before Adam sinned that he would sin. This is the inconsistency between God and free will all over again.

## OTHER PROBLEMS WITH GENESIS

While Collins accepts the evolutionary development of humans from earlier forms, he still apparently believes in acts of special creation when it comes to human beings. After all, Collins asserts that humans were the result of a special divine effort at fine-tuning the physical constants of the universe; that they are the intended goal of evolution; that they have supernatural faculties, like free will and innate morality; that they have an immortal soul and will experience an after-life; and that only humans enjoy a fellowship with God.

An additional problem is the unbelievable ages of Adam's descendants. Methuselah was 969 years old. Noah was 500 years old. Speaking of Noah, what is Collins's interpretation of the Flood story? Collins does not address the many contradictions in the story, including the scientific impossibility of saving all the millions of species known to science. I assume that as a knowledgeable scientist he also regards this as an allegorical morality lesson, but exactly what we learn from it, if anything, is open to debate. The vast humanity-destroying flood is a great story retold from the Mesopotamian epic poem *Gilgamesh*.[45] It is a tribute to the human imagination but indefensible as history, science, or evidence of a loving God who doesn't make mistakes.

In seeking to explain the world, myths and religion are just bad science. The story of the Tower of Babel and the proposal that "the whole earth was of one language" is inconsistent with anthropologic, archeological, and linguistic studies. The biblical story of Sodom and Gomorrah is another difficult story to reconcile with a just God. When the evil, lustful homosexuals of the city demanded access to two angels (young men?) who are visiting Lot, he attempts to save them by offering them his wife and two daughters "to do ye to them as is good in your eyes" (Genesis 19:8). Giving your womenfolk to be sexually abused as a bribe to spare your guests homosexual rape is a morally indefensible offer. Is Lot really the only truly righteous man Yahweh chooses to save? When God destroyed Sodom and

Gomorrah, were there no pregnant women, no blameless children, and no innocent fetuses in the city? A just God? A wrathful God? A myth or another allegory to teach another lesson?

The rest of Genesis proposes to be a history of the creation of the nation of Israel, starting with Abraham and ending with the story of Joseph in Egypt. It reads like an engrossing soap opera with heroes, antiheroes, greed, lust, courage, friendship, betrayal, and so on. The author of these chapters could well have been Niccolo Machiavelli. To pick one problematical episode, consider chapter 34. Dina, the daughter of Jacob, is raped by a regional prince, Sichem, who subsequently falls in love with her and makes a marriage proposal to Jacob. Sichem offers money, peace between the tribes, land rights, and trade. Jacob's family deceitfully responds that if Sichem and the males of his tribe are circumcised as required by God's law, they will accept his offer and allow the marriage. Sichem agrees and convinces all the males to be circumcised. Three days later, while Sichem and his men "were in pain," Jacob's sons kill them all and carry off their wealth, women, and children. Jacob's only worry is that the deceit might make him "loathsome to the inhabitants of the land" and he will be destroyed. This is an example of pragmatic politics, not moral transgression and repentance. What divine message does this chapter convey? Is the message clear and important for humankind?

The book of Exodus presents Bible literalists with many problems, not the least of which is the Passover plague story. This is the last of ten plagues Yahweh inflicts on the Egyptian people because of the hard heart of the pharaoh who refuses to let Moses and the Jews leave Egypt. God tells Moses to have the Jews take the blood of a one-year-old male lamb that is without blemish and was slaughtered during twilight and use it to coat the doorposts and lintels of their homes. (*Why Yahweh cannot tell Egyptian babies from Jewish babies without this gruesome guide is not explained.*) Yahweh then kills the first-born in every house not so marked. (*Oops! Dummy—that was a female lamb.*) Why would a loving, compassionate, and just God do such a horrible thing and cause so much suffering? Were the Egyptian people all as hard-hearted as the pharaoh? What evil had they done to deserve such universal punishment? Why was Yahweh so concerned with the details of Passover that he divinely inspired some unknown author to write and preserve them?

What does Collins think of Exodus and the rest of the Pentateuch, which gives the Mosaic Law to the Jewish people? Poetry or historical fact? Literal or metaphorical interpretation? The "Historical Books," Joshua, Judges, and so on, and the "Prophetic Books" seem to be concerned with political infighting for control of people and land. Their authors use Yahweh as a means of justifying their actions and debates on whether or not the Jewish people are meeting their commitments in the covenant with their God. How strong is the evidence, in Collins's opinion, that clear, verifiable prophecies, properly made before the event, came true? How many did not come true? Is the so-called fulfillment of prophecies clear and strong evidence of the divine inspiration of the Bible? Is the fulfillment of the prophecies of the famous French astrologist Michel de Nostredame, known as Nostradamus, and the modern TV celebrities who predict the future also evidence of divine inspiration?

## THE BIBLE AS LEGEND

The Bible is a crucial part of the thinking that led Collins to take the gigantic leap from a belief in a god that wrote Moral Law into human nature to the divinity of Jesus Christ and the specific dogmas of Christianity, such as the Trinity and original sin. The accuracy of the Bible, the only support for the personal God, Jesus, is an area of disagreement between us. Collins's acceptance of the biblical account as evidence for the divinity of Jesus was in part based on his reading of C. S. Lewis. The paragraph from Lewis that convinced Collins was:

> I am trying here to prevent anyone saying the really foolish thing that people often say about Him: "I'm ready to accept Jesus as a great moral teacher, but I don't accept His claim to be God." That

> is one thing we must not say. A man who was merely a man and said the sort of things Jesus said would not be a great moral teacher. He would either be a lunatic—on a level with a man who says He is a poached egg—or else He would be the Devil of Hell. You must make your choice. Either this man was and is the Son of God or else a madman or something worse. You can shut Him up for a fool, you can spit at Him and kill Him as a demon; or you can fall at His feet and call Him Lord and God. But let us not come with any patronizing nonsense about His being a great human teacher. He has not left that open to us. He did not intend to.[46]

Lewis asks, was Jesus liar, lunatic, or Lord? I do not regard Jesus as liar, lunatic, or Lord but as a human legend. There is sufficient evidence to accept that Jesus wandered about Palestine preaching a new theology that attracted followers. These followers told others about Jesus, about what they thought they heard him say, and some of the "wonders" he performed. Just like the legends of Paul Bunyan or George Washington, these stories in retelling over time became more detailed, more interesting, and wonderful. As we discussed earlier, this process can be seen in the synoptic Gospels. Mark's description of Jesus and the events of his life is the simplest and least detailed. Matthew and Luke add more miracles and more detailed quotations. The birth of Jesus is not covered in Mark or John. Matthew covers it in eight verses and Luke in thirty-nine. The Resurrection story takes eight verses in Mark. Matthew adds details in twenty verses and Luke expands the story to fifty-three. John uses fifty-six. As we have demonstrated, this has all the earmarks of a legend in the making. The final result is to make a god out of a holy man.

C. S. Lewis's argument in favor of the divinity of Jesus depends on our "knowing" the sort of things Jesus said. But, we do not know exactly what, if anything, he said about being the Son of God. If he did say it, what exactly did he mean? If Jesus claimed to be divine or even simply a messenger from the divine, he would not be the first or the last human to have such a mistaken belief. If the apostles were in hiding during the Crucifixion, and the women were "looking on from afar," how do we know exactly what Jesus said on the cross? When Jesus was alone in the wilderness and tempted by the devil, who was there to record the dialogue we find in Mark 3 and Luke 4? So, arguing literally from what Jesus said as written in these texts is not compelling since we really do not have exact quotations.[47]

I regard Jesus as a teacher and interpreter of the Jewish law and moral behavior, but in every sense a man and only a man. Collins accepts the biblical claims and stories as spiritual truth, not scientific truth or rational proof of Christianity. When he was " born again" in front of that frozen waterfall in the Cascade Mountains, it was a real and powerful experience—but an emotional, not a rational one. Collins admits as much: "On a beautiful fall day, as I was hiking in the Cascade Mountains during my first trip west of the Mississippi, the majesty and beauty of God's creation overwhelmed my resistance. As I rounded a corner and saw a beautiful and unexpected frozen waterfall hundreds of feet high, I knew the search was over. The next morning, I knelt in the dewy grass as the sun rose and surrendered to Jesus Christ."[48] Living in California, I have been fortunate to have had firsthand experience with the majesty and beauty of the natural universe—the restless waves pounding Seal Rock in San Francisco Bay, the inspiring redwoods, the lava beds of Lassen Volcanic Park, the wonderful vistas of Yosemite, the barren beauty of Death Valley. Viewing these natural wonders, I have felt an emotional reaction that combines pleasure, awe, and contemplative introspection in these experiences, but they are all the result of very powerful and complex natural forces and activities. God is not self-evident.

## ARE ATHEISTS THE REAL BIBLE LITERALISTS?

Some believers who are not Bible literalists might object to my analysis and use of Bible quotations as nitpicking. They might claim that my critique is based on legalistic details and that I have missed the broad, overall message or spirit of the scriptures. John Haught, a theologian at Georgetown University, in his recent book *God and the New Atheism: A Critical Response to Dawkins, Harris, and Hitchens*, admits, "Reading certain passages of the Bible, including the Christian Scriptures, can be a dangerous and bewildering experience if one has not first gained some sense of the Bible's overarching themes."[49] Like many theologians, Haught seems to feel that we should leave biblical interpretation to be debated in the rarified atmosphere of theology departments.

I am aware that some theologians claim that even if the Bible were not divinely inspired, and even if it contained factual errors and contradictions, it could still be used to gain spiritual insights, namely, knowledge about the supernatural, if only I could read it with the same preconception about the reality of the supernatural on which theology is based. I acknowledge that the development of Christianity, based on the Bible, was unquestionably an important influence on Western civilization. While Christianity did contribute to much human suffering, it was a generally positive move toward a more humane world. It was by no means the only humanizing or civilizing factor. This positive influence, however, is not evidence of divine inspiration or the absence of error, exaggeration, and bias in the biblical writings.

Those who regard the Bible as a divinely inspired message and general moral guide, but not a word-for-word reproduction of Jesus' sayings and teaching, may feel that citations of contradictory or no-longer-accepted moral practices or even outright errors do not discredit it. This view of the Bible would invite us to think that God's revelation to a sinful humankind is simply the general message, "Be good; don't do evil." If this were all that the Gospels were supposed to be, who could disagree? Isn't the real question that we want revelation to answer, "What is good behavior and what is evil behavior?" Isn't the Bible supposed to be God's detailed plan for humankind and a moral code that tells us what behaviors are good and what behaviors are evil? As is constantly observed, the devil is in the details. Human laws do not provide general direction like "It is illegal to kill." The law includes details that distinguish between justifiable homicide, manslaughter, and three degrees of homicide. The law answers specific questions like "Can we kill animals?" "Can we kill in self-defense?" and "Can we kill ourselves?"

If you argue that we should not pay too much attention to the biblical details, such as observance of the Sabbath, for example, are there any specific commandments that we must observe as essential to the Gospel message? How do we determine which details or specific language are essential and which we can ignore or interpret as allegory or legend? The trouble with treating the texts symbolically or metaphorically is that there are many interpretations of symbols and metaphors and no criteria for determining which of them was intended by the author.

There are some liberal theologians who no longer believe in the virgin birth, miracles, or the Resurrection, but this is not what Collins describes as his faith. His faith requires and depends on the Bible being God's self-disclosure to humankind through the person of Jesus. Collins sees the Bible as a sacred book, not as the efforts of some first-century Jews trying to persuade their fellows to accept their version of Jesus' life and teaching. Christianity is a religion of love and compassion but it is also a religion of doctrines and dogma. Theologians read and analyze the scriptures and writings of the apostolic fathers in detail, parsing the meaning of words and sentences to support or refute the doctrines of the Fall, the Virgin Birth, the Incarnation, the Resurrection, and the Trinity. These theologians all have blind faith in the existence of God and the supernatural, and within that framework all kinds of "rational arguments" and scholarly conclusions are possible. They assume without good and compelling evidence that the writings are divine or contain insights into the divine.

If you want to cherry pick what is and is not fact in the Bible, what are the rules? If you do not believe biblical details are important, why do you think God authorized their inclusion in the Bible? For example, can you see why God thought it was important to include chapter 1 in the first book of Chronicles? Do you feel that anyone reading chapter 2 of the book of Revelation would get a clear idea of what Jesus was trying to say through John the Divine? Would Jesus' message be incomplete if these chapters were omitted from the Bible? Is the information in every chapter important to your understanding of God and what is needed for salvation? The passages I have cited exemplify many passages that illustrate some of the multiple problems inherent in the Bible. It is the text of the Bible itself that forces me to conclude that it cannot be used as historical or scientific evidence in support of the divinity of Jesus Christ. I would not question the word of God, but I have good reasons to believe the Bible is not the word of God and no evidence that it is an authentic divine revelation.

## WHAT DOES THE POPULARITY OF THE BIBLE PROVE?

What accounts for the popularity and wide dissemination of the Bible? For thousands of years after Constantine made Christianity the state religion of the Roman Empire, only the authorized texts of the Bible were allowed to be read or circulated. The Bible had no competition. Before Johannes Gutenberg invented mechanized printing in 1440, the entire Western world was centrally organized around a single hierarchal, politically powerful Holy Roman Catholic Church. Cardinals, bishops, and priests had influence everywhere from the royal courts to the smallest village. People spent an extraordinary amount of time in prayer and religious rituals. Art was full of Christian imagery and great cathedrals were built. Is it any wonder the Bible was the first book printed, and that so many were printed and made available? Handwritten books were expensive and only the clergy or the wealthy owned them. If you could afford only one book, it would be the Bible. What better way to establish in your own mind or in the view of the omnipresent clergy your devotion to Christ and his church? The Protestant Reformation, with its emphasis on individual interpretation and direct access to the Bible, promoted further translation and dissemination.

Many believers support their argument that the Bible is the word of God by citing statistics on the large numbers of copies and translations existent. But they draw a false conclusion by claiming that a text's popularity says anything about the truth of its contents. Islam is currently the fastest-growing religion. What will happen to this claim that the popularity of the Bible verifies the truth of its contents when the number of Qur'ans exceeds the number of Bibles? In contrast to Christians, who own but rarely read the Bible, Muslims read the Qur'an daily.

The American Bible Society estimates that the Bible is accessible to 98 percent of the world's population. If true, and the Bible is the clear and persuasive word of God, why isn't the majority of the world's population Christian? If you were God and you wanted to intervene in human history to communicate an important message to all of humankind, why would you write a book in Hebrew and Greek that favors this small geographic area and ignore for centuries the peoples of Africa and Asia?

## NOTES

1. George Gallup Jr., *The Role of the Bible in American Society* (Princeton, NJ: Princeton University Press, 1990).
2. H. M. Teeple, *The Historical Approach to the Bible* (Evanston, IL: Religion and Ethics Institute, 1983).
3. Noncanonical books referenced in the Bible: http://en.wikipedia.org/wiki/Lost_books_of_the_Old_Testament.
4. J. Pelikan, *Whose Bible Is It?* (New York: Viking Penguin, 2005).
5. E. Pagels, *The Gnostic Gospels* (New York: Random House, 2004).
6. Collins, *Language of God*, p. 53.
7. R. Price and J. Louder, eds., *The Empty Tomb: Jesus beyond the Grave* (Amherst, NY: Prometheus Books, 2005), pp. 435–52.
8. Collins, *Language of God*, p. 223.

9. F. F. Bruce, *The New Testament Documents: Are They Reliable?* 6th ed. (Grand Rapids, MI: William B. Eerdmans, 1981).

10. C. Blomberg, *The Historical Reliability of the Gospels* (Powers Grove, IL: Intervarsity Press, 1987); G. Habermas, *The Historical Jesus* (Joplin, MO: College Press, 1996).

11. Blomberg, *Historical Reliability*, p. 11.

12. Ibid., p. 68.

13. Ibid., p. 152.

14. Habermas, *Historical Jesus*, p. 11.

15. Ibid., p. 262.

16. L. Strobel, *The Case for Christ* (Grand Rapids, MI: Zondervan, 1998).

17. Habermas, *Historical Jesus*, p. 228.

18. G. Ludeman, *Jesus after 2,000 Years* (Amherst, NY: Prometheus Books, 2001).

19. Habermas, *Historical Jesus*, p. 273.

20. R. M. Helms, *The Bible against Itself* (Altadena, CA: Millennium Press, 2006).

21. B. F. Akerley, *The X-Rated Bible* (Venice, CA: Feral House, 1998).

22. B. D. Ehrman, *The Lost Christianities: The Battle for Scripture and the Faiths We Never Knew* (New York: Oxford University Press, 2003).

23. Gallup Poll: http://www.gallup.com/poll/22885/TwentyEight-Percent-Believe-Bible-Actual-Word-God.aspx.

24. International Social Survey Program: Religion 1993. National Opinion Research Center at University of Chicago, as obtained by Inter-University Consortium for Political and Social Research, University of Michigan, in G. Bishop, "What Americans Really Believe," *Free Inquiry* 19 no. 3 (Summer 1999).

25. Gallup Poll: Religion. http://www.gallup.com/poll/1690/Religion.aspx (accessed November 11, 2008).

26. Gallup Poll: http://www.gallup.com/27682/OneThird-Americans-Believe-Bible-Literally-True.aspx (accessed November 12, 2008).

27. Collins, *Language of God*, p. 83.

28. Ibid., p. 228.

29. Ibid., p. 209.

30. Ibid., p. 153.

31. Ibid., p. 158.

32. Ibid., p. 175.

33. *The Complete C. S. Lewis* (San Francisco: Signature Classics Harper-Collins, 2002).

34. Ibid., p. 593.

35. Ibid., p. 594.

36. Ibid.

37. Ibid., p. 596.

38. Ibid., p. 595.

39. Ibid.

40. Ibid., p. 599.

41. Ibid., p. 600.

42. Ibid., p. 599.

43. Ibid., p. 590.

44. Ibid., p. 595.

45. D. Damrosch, *The Buried Book: The Loss and Rediscovery of the Great Epic of Gilgamesh* (New York: Henry Holt, 2007).

46. *Complete C. S. Lewis*, p. 36.

47. B. D. Ehrman, *Misquoting Jesus: The Story behind Who Changed the Bible and Why* (San Francisco: Harper, 2005).

48. Collins, *Language of God*, p. 225.

49. J. Haught, *God and the New Atheism: A Critical Response to Dawkins, Harris, and Hitchens* (Louisville, KY: Westminister John Knox Press, 2008), p. 100.

## *Chapter* 7

# NATURALISM (ATHEISM AND AGNOSTICISM)

> *I have recently been examining all the known superstitions of the world and do not find in our particular superstition (Christianity) one redeeming feature. They are all alike, founded upon fables and mythologies.*
>
> —Thomas Jefferson, personal letter to Dr. Woods

## COLLINS AGAINST ATHEISM

Collins approvingly quotes Galileo's affirmation, "I do not feel obligated to believe that the same god who has endowed us with sense, reason, and intellect has intended us to forgo their use."[1] As a biological scientist, Collins's faith is most threatened by evolution, so he explores four possible options people use to harmonize religious beliefs with this scientific concept. These are (1) atheism or agnosticism, (2) creationism, (3) intelligent design, and (4) *biologos*, Collins's term for the acceptance of both scientific evolution and scripture-based faith.

I agree with Collins that, at this point in history, evolution is a battleground for religion and that believers who try to find scientific justification for the truth of their beliefs are ultimately doomed to fail. Collins devotes two chapters to refuting and answering critiques of evolutionary science by both young-earth creationists and intelligent designers. It is interesting to note that in refuting intelligent design, Collins says, "ID [intelligent design] is a 'God of the gaps' theory, inserting a supposition of the need for supernatural intervention in places that its proponents claim science cannot explain."[2] This is exactly what Collins does to explain the big bang, the anthropic coincidences, and the Moral Law. While I believe there are more and better arguments to use, Collins does a respectable job. Science and the US courts have defined these proposals as religious, not scientific, interpretations. I, therefore, pass over Collins's chapters arguing against creationism and intelligent design and compare his first and fourth options, namely, scientific atheism and theistic evolution, or as Collins has renamed it, *biologos*. I will discuss whether Collins achieves his objective of the production of a satisfying synthesis between the Christian worldview and that of evolutionary science and science in general.

The first option we will consider is atheism or agnosticism. Collins divides atheism into two types, "weak," the absence of belief in God, and "strong," the conviction God does not exist, but then he decides to address only the strong version. Collins dates atheism back to the eighteenth-century Enlightenment, although it has existed in one form or another since the concept of supernatural beings was proposed. The Greek philosophers Democritus and Epicurus argued against theism. The fact that atheists, people who denied the god of their culture, have been universally persecuted, punished, and killed by theists obviously limits its popularity as a worldview. Collins attributes the growth of atheism to rebellion against the oppressive alliance of the Roman Catholic Church and the royal state governments that accumulated wealth and power at the expense of the common man during the sixteenth, seventeenth, and eighteenth centuries. This is an accurate but incomplete analysis. The Enlightenment was also characterized by a breaking away from dogmatic ways of thinking and a rejection of the uniform answers to questions about the nature and purposes of human life. The Enlightenment was not only a political revolution but also an intellectual one in which reason rather than authority guided the examination and interpretation of the natural world. It was during this period that science became a discipline separate from philosophy and theology.

The growth of atheism roughly coincides with the expansion of science. The philosophical wars, which preceded the science wars, involved proofs and critiques of proofs of God's existence using logic and reason alone. Intellectuals such as Anselm, St. Thomas Aquinas, Descartes, Spinoza, Leibniz, Locke, Berkeley, Hume, and Voltaire, to mention only a few, were the contestants. Philosophers such as d'Holbach, d'Alembert, and La Mettrie offered the most compelling arguments challenging the proofs of the existence of God and affirming the irrationality of the existence of any kind of god. Famous philosophers such as Kant, Hume, and Kierkegaard expressed the consensus of modern philosophers that there is no logically compelling proof of the existence of God. However, Collins selects Sigmund Freud, E. O. Wilson, and, most especially, Richard Dawkins as his representatives of atheism. By choosing these men, Collins, in a sense, does what he accuses Dawkins of doing: setting up straw men.

## COLLINS VERSUS FREUD

Freud was a medical doctor (not a philosopher) who described belief in God as wish fulfillment, a topic I discussed in chapter 3. Freud was attempting to explain why people choose to believe in God and only indirectly addressed the truth or falsity of God's existence. There is some truth to describing religious faith as wish fulfillment; however, the argument for atheism does not depend on the validity of Freud's psychoanalytical theories.

I was enrolled in psychiatry during all four years of medical training at UCLA. I was very open-minded, and, in fact, eager to learn about why we behave the way we do. But I soon determined that all the psychiatrists were doing was merely explaining behavior after it happened, just fitting fact to theory. They were not able to accurately predict behavior, nor were they very effective in altering undesirable behavior in controlled studies. I could appreciate that discussing problems with a nonjudgmental person can have beneficial effects, but it was also clear that the treatment of mental illness was substantially improved by psychotropic medications.

I need not defend Freud's psychiatric theories as substantive to the main arguments for the truth of atheism. In keeping with Collins's selection of Freud as the spokesperson for atheism and C. S. Lewis as the champion of "rational" Christianity, Collins points to the fictional debate constructed by Harvard psychoanalyst Armand M. Nicholi Jr. in his book *The Question of God.*[3] Collins uses this debate to make the point that the God people wish for is not at all like the "real" God as described by Lewis.

Humans would like to believe in a kind, compassionate, understanding, loving, forgiving, protective, nurturing father figure. However, according to Lewis, God's nature demands that humans demonstrate constant, voluntary obedience to God's Moral Law or face the eternal torments of hell to satisfy God's justice. Lewis correctly distinguishes kindness from love but claims God's love does not include kindness. I cannot accept this claim. Real love, and certainly perfect love, must include kindness. While it is true that humans can be unkind to someone they "love," human love is conditional and sometimes selfish and imperfect. God could be kind to someone God did not love, but God could not be unkind to someone God loved perfectly. Lewis's attempt to redefine the nature of love as it might apply to God is an unintelligible distortion and describes a God undeserving of worship. Wish fulfillment does not explain God, or explain away God, regardless of the kinds of gods humans conceive of or wish for.

## COLLINS VERSUS DAWKINS

We next consider Collins's characterization of the arguments in support of atheism of evolutionary biologist Richard Dawkins. First, Collins says Dawkins argues that there is no need to resort to God to explain "biological complexity and the origins of human kind." This is correct. Science can do a very reasonable, coherent, and rational job of explaining both biology and the cosmos in terms of completely natural causes. Dawkins points out that we don't need supernatural concepts to make sense of the universe. Even though Dawkins reacts with unrestrained hostility to those who do not accept this idea, the reasonableness or truth of his main arguments are not diminished. It is also true that the success of scientific explanations of the universe have rendered supernatural ones obsolete. Collins endorses scientific explanations of the natural world but claims that science neither proves nor disproves the existence of the supernatural. If no scientific observation of any phenomena could prove the existence of God or a supernatural reality, then science is irrelevant to theism and theism is irrelevant to science. Collins simply believes in the theory of the supernatural. However, when you leap unaided by science to a belief in a supernatural reality and acceptance of a personal God like Jesus, Allah, or Yahweh, who specifically creates humans, has a providential plan for each of them, hears and grants intercessory prayer, and requires specific behaviors from them, one can expect to arouse emotional as well as intellectual responses from militant atheists like Dawkins.

Dawkins's second argument is that religious faith is "blind trust in the absence of evidence even in the teeth of evidence."[4] This is an apt description of Collins's faith despite his claims of rationality. The so-called evidence for the existence of God as a supernatural being (the Moral Law, the big bang, and the anthropic coincidences) might have led Collins to theism, but it does not support any specific religion, namely, Christianity. More and different evidence is needed to establish the validity of the personal God unique to a given religion. If you can prove that a personal God of one religion exists, for example, Christianity, you have simultaneously disproved the existence of personal Gods of all other religions, for example, Islam. Collins's book provides no such evidence of the divinity of Jesus other than the biased stories in the Bible. Collins's response to the charge of irrationality is to claim religious belief is at least plausible. Yes, I agree—but only if by "plausible" he means possible but extremely unlikely.

Dawkins's third argument for atheism, according to Collins, is the great harm done by believers, that is, the results of believing in God are not universally good or desirable. Collins acknowledges that great harms have occurred in the name of religion. He attempts to mitigate this admission with two invalid arguments: first, that evildoers are not "true" believers, which depends on who gets to define "true" or orthodox belief; second, that the evil is balanced by great acts of compassion. But neither of these arguments eliminates the evil nor justifies it.

Collins cites examples of the "wonderful things" done in the name of religion: Moses freeing the Jews from Egyptian-imposed slavery (for which there is no historical evidence); William Wilberforce and the abolitionists ending slavery in Great Britain; and Martin Luther King Jr.'s civil rights movement. If the Bible is historically accurate, Moses and the Jews used their freedom to commit genocide against the inhabitants of Palestine and established their own slaves. There were Christians on both sides of the English slavery debate. There were many secularists who supported abolition; the movement was not uniquely religious. Likewise, Dr. King was a religious person and this was a big part of his motivation. But the civil rights movement was a social movement supported by liberal secularists and individuals of different religiosity. In fact, the main opponents of racial equality were Bible-thumping conservative Christians who immediately set up private schools for whites and provided second-rate education to African Americans. I acknowledge that Christians have provided compassion for the less fortunate, comfort for people in time of loss, and ministry to the sick, but all these good works are possible without adding the religious element.

Good works performed by believers do not justify the evil committed by believers in the name of God, evil in their minds justified by their Christian beliefs. Has the religious evil been necessary to obtain the religious good? Would great acts of compassion occur without the evil? What difference does religion make if atheists and believers both do evil and both exhibit virtue and good behavior?

This argument, that religion is totally evil in its effects on society, is again a straw man that Collins knocks down. I agree with him that religion and theism are not all bad, but this is not a contribution to the argument about whether the atheists' denial of the existence of a god, personal or impersonal, is true or false.

Finally, Collins attacks Dawkins's argument that the Moral Law is not a divine insertion into human behavior but can be explained by the interaction of complex genetic mechanisms and the environment acting in accordance with natural laws that are blind and indifferent to human purpose. Dawkins proposes that the evolution of advanced brain function, that is, the mind, may allow humans to rebel against nature's indifferent drive for reproductive success and survival and impose an "unnatural," disinterested altruism to achieve unique human purposes.

Just because Dawkins accepts that altruistic behavior exists and that it exhibits humans' natural tendency to form opinions about the desirability of different behaviors, a process that Collins labels the Moral Law, does not mean that Dawkins is contradicting his atheistic convictions. Dawkins is still denying that altruism or any other behavior comes from God. While Dawkins characterizes the laws of nature as "blind, pitiless," and "indifferent," this does not mean that he believes all the products of these law are also blind, pitiless, and indifferent. So, I cannot understand how Collins can defend his statement that Dawkins's acceptance of altruistic behavior is somehow a paradox that discredits Dawkins's atheism.

Collins's most plausible criticism of Dawkins is that Dawkins's claim that science demands one to accept atheism goes beyond the evidence. This is not the universal opinion of atheists. Science, while finding no evidence of the supernatural, does not demand atheism. A scientist could rationally be an agnostic or a deist. It is the belief in a personal God that is irrational and unscientific. Scientific evidence is, however, more consistent with the rationale and arguments of atheism than those of the theists.

A person who says *A* causes *B*, and then *B* causes *C*, and then "God" comes along and causes *D*, then *D* causes *E*, and so on, would not be classified as a scientist. Any mathematical proof can be "demonstrated" if you allow the mathematician to arbitrarily insert a "God fudge factor" or constant into his equations. Evolutionary biologist and popular author Stephen Jay Gould says, in the quotation used by Collins, "Science can work only with naturalistic explanations, it can neither affirm nor deny other types of actors (like God)."[5] However, scientists as rational beings cannot be prohibited from drawing philosophical conclusions from our evidence. Any human can definitely comment on the logical impossibility of God and the lack of any evidence for a supernatural reality.

Scientists cannot, as Gould suggests, forget logic and philosophy, which render the idea of God illogical and irrational. I agree with Collins that the issue is not resolved by claiming that science and religion are two separate, nonoverlapping methods of acquiring truth about the universe and our place in it. Science deals with reality, and religion with supernatural unreality.

## THE CASE FOR ATHEISM

I find Collins's critique of atheism very superficial and unconvincing. He, apparently, passively accepted the atheism of his family and did not conduct a detailed investigation of the arguments and evidence supporting it. If you want to understand the atheist worldview, I suggest you read books written by atheists such as *Atheism: The Case against God*, *The Atheist Universe*, or *Atheism Explained* as a start.[6] Collins has not addressed the main arguments and evidence that have persuaded me that atheism is the most believable of Collins's options.

So, if you will bear with me, let me review what I would consider the best arguments and evidence for atheism. There are rational arguments that lead to the conclusion that God does not exist. While there are various definitions of God, for purposes of argument "God" will be defined as a being that is perfect in all aspects. This is a fair definition of the God Collins worships and defends. This means the being possesses perfect, errorless knowledge and has no needs that must be satisfied, no imperfections that have to be corrected. The being is absolutely good and has nothing evil in its nature. This is the nature of the God of the ethical monotheistic religions Christianity, Judaism, and Islam. Using these ground rules, let us begin.

## IS OUR UNIVERSE THE WORK OF GOD?

"Why is there something rather than nothing?" There has got to be a creator God, right? The concept of nothingness is unintelligible. There have to be either natural, material things or beings, or supernatural, immaterial things or beings, to even initiate a reasonable argument or discussion. In other words, there has to be something to talk about. We know there are natural, material objects, a real universe of which we are a part. We can only believe, speculate, and imagine a possible supernatural existence.

According to the teaching and sacred texts of most religions, God created the universe out of nothing. At some point the universe came into existence because God willed it into existence. Before that point there was nothing. Nothing except God. Before creation, all that existed was perfect goodness, that is, God. There was no trace of evil. Before that point there was nothing that was imperfect. Before that point God was perfectly satisfied with his own perfect being or nature and needed nothing to improve God's unchanging nature. God had no unfulfilled desire. A God that was all good and completely perfect and satisfied would have no need to create something containing imperfection and evil from nothingness. There is a universe, however, that is, despite its beauty and pleasures, imperfect and contains evil, suffering, and injustice. The fact that something such as this imperfect universe exists is not consistent with our definition of God. A perfect God would, by his nature, not be a creator God. Our clear, direct, personal experience with evil and imperfection in the natural universe makes God's existence logically impossible.

Believers respond that there may have been no need for God to create such a universe, but he wanted to do it this way, and who are we to limit the will or whim of God? If we accept this proposed explanation, we have to accept either a God with motivations and needs to satisfy, like us imperfect humans, or a God that is unintelligible, not rationally motivated, mysterious, and incomprehensible. Neither option is a logical or acceptable description of God.

Is God capricious and arbitrary? Wouldn't this make God a sadistic manipulator of his creatures, giving them rationality and intelligence that are faulty and of no use in determining their creator's purposes? Doesn't this defense of God depict a God who wants, creates, or is powerless to prevent evil? This explanation would allow for the existence of an impersonal God that is unconcerned with human concepts of good and evil and oblivious to human suffering and prayers. Is this a creator God worthy of our worship?

## IS GOD EVIL?

To refute this argument that a perfect God could not create an imperfect universe, you could claim along with the German philosopher mathematician Gottfried Wilhelm Leibniz (1646–1716) that God created the best of all possible universes. This contradicts the facts of our experience. To claim this is the best universe possible, you have to deny the existence of imperfection, evil, suffering, and injustice. Since we are seeking truth as human beings, we have no practical alternative but to use the human context to assign values. So, unless we want to abandon all meaning and render rational debate impossible, we have to admit that human beings universally experience imperfect, painful, unjust, and evil behaviors and events.

Humans both cause and experience physical and mental pain. Evil exists; it cannot be denied. Evil is not just the absence of good, like darkness is the absence of light. It exists as an alternative to good. God could have created a universe without darkness, a universe eternally bathed in light. One phenomenon or concept does not necessarily imply the existence of the other. Creation of a perfect universe without evil is not a logical impossibility like creating a circle that has three sides.

Believers are wrong when they claim that without evil there is no good, so the best of all possible worlds must contain evil. How do believers logically and rationally defend this concept of the necessity of evil without accepting the result that God's nature cannot be all good and must contain evil? How do they explain that, according to their belief system, God actually did create a universe that has no evil? They call it heaven. Compared to heaven our universe is clearly not the best universe God could create.

Moreover, for biblical literalists, the original universe, according to Genesis, was "good." Adam and Eve were created sinless. There was no evil until Eve bit the apple. So, it was not beyond God's power to create a universe that had no evil in it. What necessitates an intelligent, all-powerful God to create an imperfect universe full of man-made and God-made, that is, natural, evil as a way station for humans before they arrive at the state of eternal, blissful enjoyment of pure, loving goodness? The angels apparently do not go through such a trial of suffering. It does not seem to me that this restriction on God's power is reasonable or consistent with the definition of a God worthy of worship. So Leibniz was wrong; this is not the best of all possible worlds and was not created by God.

## THE THEODICY

The most powerful and persuasive argument for the nonexistence of an ethical, personal God like Jesus, Allah, or Yahweh is called the *theodicy*. This term, which was created by Leibniz, comes from the Greek words for God's justice and refers to the age-old problem of reconciling belief in a good, loving, and all-powerful God, and the existence of the pain and suffering of his innocent creatures. No other problem has been as responsible for unbelief and loss of belief in a personal God as this problem.

There is a large body of literature on the topic in philosophy, theology, and religion, spanning thousands of years and involving some of the best and brightest humans ever to have existed. Just as no one has produced a convincing logical proof of God's existence, no one has devised a convincing, logical answer to reconcile the irreconcilable. The following are the main lines of arguments used in the vain attempt to preserve the notion of a nonevil, all-powerful personal God.

## THE FREE WILL DEFENSE

Early on in this theodicy debate, evil was separated into two kinds: moral evil caused by humans, for example, sin, murder, lying; and natural evil, namely, acts of God or "Mother Nature," including earthquakes, floods, and so on. The classic response to human-caused evil is that God gave humans the gift of free will. So, let me briefly address what I believe to be the flaws in this defense. There is a growing body of literature by scientists and philosophers that questions the existence of free will. A natural explanation of the human mind and consciousness is one of the ongoing and major uncompleted goals of science. It is highly probable that, in the not too distant future, neurocognitive science will be able to show that free will is a powerful illusion of the human mind. However, for the purpose of argument, I will accept that humans have free will. I will also argue that the existence of free will leads to the conclusion that a personal God does not exist.

Again, we go back to our working definition of a personal God. This God is perfectly good, loving, and all-powerful. Let's say we agree with Collins that God used an evolutionary process to create humans. Beginning with the big bang, God continued the process until the point when God was ready to introduce the immortal soul and the Moral Law and create literally a new and special species of *Homo*. Would God also incorporate the faculty of free will?

Until this point an unbroken chain of cause and effect controlled natural evolution. The natural laws are uniform, or at least do not change over vast periods of time. This renders events predictable once their causes are discovered. Suddenly, God gives humans the capacity to produce an event without a predictable cause. If you deny this, then you are denying the existence of free will. To say that every event has sufficient and necessary cause means that every event is completely determined by the preceding set of natural circumstances.

To make this clearer, consider an example. Instead of driving your car home from work, you take a taxi. What caused you to take a taxi? You received a credible threat that your car was going to be bombed. If you feel that this threat caused you to take the cab, then the threat created such a strong motivation that driving your car became impossible. Your action becomes 100 percent determined by events outside of you and 100 percent predictable. Whatever caused you to take a taxi was strong enough to determine your action; therefore, your will is not free. Simply saying, after taking the taxi, you "feel" you could have done otherwise does not prove that, in fact, you could have driven your car or that the threat was not a real cause for your action. This is determinism, not free will. If you drove your car home because your strongest motivation was a desire to show that you are fearless, your action was still determined by your strongest desire and you were not free to do otherwise. If we had sufficient information to determine your detailed genetic makeup and behavioral tendencies, we could predict which action you would take.

To have truly free will is to be able to break this chain of cause and effect and do something totally unpredictable, without strong reasons or causes, including the strong conscious motivation to do something unpredictable, which would then be a sufficient and necessary cause of your "unpredictable" action. Your actions would not be enforced by nature. They would not be the necessary results of previous events and circumstances. No motivation or circumstance, no matter how strong, could force your action. These free actions are random, unpredictable, not caused by God. In fact, they are uncaused. This is the kind of free will that God is supposed to have given as a gift to humans.

So, believing in free will means giving up the chain of causation. You would have to claim that you could freely choose to take some action in an unpredictable way, without any good reason or any reason (cause) at all. Once you say that you took a certain action because of something that motivated you or gave you a reason to act, that something becomes the cause, the chain is restored, and you lose free will.

Now, let us return to the decision point at which God decides to insert free will into human nature. Being all knowing, God knows what the result will be. God knows humans will use free will to cause other humans to suffer. Free will is not a gift that a loving God would want to give to his creatures any more than a loving father would give a loaded machine gun to his toddler. Both would know that the gift is dangerous and inevitably would be misused and cause harm and suffering. If God proceeds to give the gift, then God has directly or indirectly created sin and suffering, or at least has permitted them when he could have prevented them. The conclusion is inescapable. God is responsible for sin and evil and is not all good and loving. It is only after God creates free will that man-made evil can enter the world.

Others say God only permits us to exercise free will because God cannot do otherwise. Really? What law requires that any of God's creatures have to have free will? We do not attribute free will to lower creatures and certainly not to inanimate creation. To suggest this solution leads to a denial that God is all powerful and can do anything that is logically possible.

Does God have the power to make humans with free will but restrict their choices to good, better, and best? A world where there are no really evil choices? Would that be a better world?

A variation on this theme is that God is using evil to teach us a lesson, to explain what he wants us to do or to clarify his message. Suffering is a poor teaching tool indeed. Do you need to burn your children in order to teach them not to touch hot objects? The arguments of the previous paragraph apply to this excuse. A really intelligent, loving teacher who is all powerful could and would have to, by virtue of that being's nature, do better. What was the lesson of Auschwitz? Was it clear? How many learned and profited from it? Was there no other way to make the lesson clear? How about 9/11 or the genocides in Rwanda or Darfur? Are we agreed on what God was teaching us? Are we willing to worship such a teacher?

Some believers say that 9/11 and other events that cause suffering are punishments from God for our sins, that is, our failure to behave as God wishes us to behave. But there would be no need to punish sin if God had not created or permitted sin in the first place. Even if you believe that justice requires punishment, it also requires that the punishment be inflicted on the evildoers, not on innocent persons, and that it be proportional to the harm done. The attacks of 9/11 probably killed some homosexuals and fornicators and maybe an atheist or two, but the majority of "sinners" were untouched. Many people whom Christians would characterize as good were killed, and much misery was inflicted on their surviving loved ones. Is this a God you would find worthy of worship and love? Is this the way an all-powerful, all-loving, all-good God would behave? Is the eternal, horrific punishment of hell required to be proportional in order to satisfy God's justice? Does the punishment fit the crime? Does God punish for the sake of punishment alone, or does God have another purpose? Couldn't an all-powerful God achieve this purpose without punishment? If God cannot act to prevent sin and human-caused suffering, how can God judge humans, who are not as powerful as God, when they do evil?

Believers often say that suffering is God's way of testing us. What kind of father would subject his children to painful tests to prove that they were obedient children or that they really loved their father? What new knowledge would an all-knowing God learn by testing that God did not previously know? Why test so many, knowing that many will fail?

Another response is that God allows evil and suffering to achieve a greater good. If believers want to use this rationale, they must respond to questions that it generates. First, no believer has produced a scholarly work demonstrating that the huge amount of suffering and evil in the universe has been, in any way, balanced by the good. To quote Christopher Hedges:

> Look just at the 1990s: 2 million dead in Afghanistan; 1.5 million dead in the Sudan; some 800,000 butchered in ninety days in Rwanda; a half million dead in Angola; a quarter of a million dead in Bosnia; 200,000 dead in Guatemala; 150,000 dead in Liberia; a quarter of a million dead in Burundi; 75,000 dead in Algeria; and untold tens of thousands lost in the border conflict between Ethiopia and Eritrea, the fighting in Colombia, the Israeli-Palestinian conflict, Chechnya, Sri Lanka, southeastern Turkey, Sierra Leone, Northern Ireland, Kosovo, and the Persian Gulf War (where perhaps as many as 35,000 Iraqi citizens were killed). In the wars of the twentieth century, not less than 62 million civilians have perished, nearly 20 million more than the 43 million military personnel killed. Praise the good and loving god![7]

Set aside whether these deaths were agonizing or instant; add to these figures the physical and mental suffering of the three to five times more civilians and soldiers who were wounded and permanently handicapped in these conflicts. Add to that the mental anguish and suffering of the families and friends who survived the physical harm to or death of their loved ones. The weight of suffering is immense. Where is the counterbalance of good that justifies this amount of pain? This argument cannot be dismissed by the observation that it is impossible to quantify suffering and good deeds and add them up because there are no units of measurement. Even if we agreed that there was more in the good column than in the suffering column, it would not justify suffering to those individuals who experienced it. If God exists, God permits this evil.

Again the question is: what are the limits of God's power? If humans can do good without evil, cannot God do likewise? If God's

Moral Law prohibits humans from using evil means to accomplish good ends, why does God have to violate the prohibition? An all-knowing God could figure out a way to do good without evil, and an all-powerful God could execute it. Why does the evil continue? Possibly because there is no God to stop it?

Finally, there is the problem of God's knowledge, which is timeless. God knows what will happen in the human time frame before it happens. God's knowledge is perfect and God never is wrong. God's knowledge is not like our knowledge, which is separate from our power to take action. Our knowing a hurricane will come does not cause the hurricane and does not give us power to stop it. But what God knows, God does. For, God's knowledge truly is power. So, before you got the telephone call, before you left for work, God knew you would take a taxi home. You could not have made God wrong by driving home. Because God's nature is unitary, God's knowledge made it absolutely necessary and certain you would take a taxi. Free will, therefore, could not exist if such a God existed. The conclusion is that you can believe in free will or you can believe in a personal God, but not both. If you agree with those who claim free will is an illusion, then you could rationally believe in a deist god, but not a personal God like Jesus, Allah, or Yahweh.

I recognize that most humans see free will as a desirable thing. However, on consideration I believe a world without free will, where we could do no evil, would be a better world. Would you rather have a world with no suffering where everyone was happy and all their desires were completely satisfied, or this world with its burden of misery, injustice, unhappiness, guilt, and free will? Do you prefer this world with free will? Then I guess you prefer this world to heaven, where happiness but not free will is available. If the souls of the departed have free will, they could do evil things in heaven. If they do not have free will, then free will is not essential to human nature.

## WHO CAUSES NATURAL EVIL?

If, after thorough investigation of the arguments against the free will defense (the problem of God's foreknowledge, the dilemma that free will proposes for causality), you accept free will as a satisfying explanation for human-caused human suffering, you still have to explain the existence of the suffering caused by natural disasters, disease, cancer, and so on due to the actions of God, not Mother Nature. These actions are unpredictable, unpreventable, and cannot be controlled by human free will. Natural disasters and disease have caused undeserved suffering for billions of innocent, decent human beings. Some would challenge the adjective "undeserved." Some Christian sects believe even embryos are contaminated by original sin, that human nature is so debased, despicable, and corrupted that none of us deserve salvation or protection from suffering. It is only because of God's love and mercy that a few are spared from the torments of hell. I cannot reconcile this view with the concept of a loving God. I could not worship a deity with this idea of justice, forgiveness, and mercy.

One resolution to the problem of natural evils that believers developed was the creation of Satan.[8] If there is evil in the universe and if God is not responsible, then someone else is. Satan could be the cause of natural diseases, HIV, cancer, and so on, which are not the result of free human actions. But this doesn't seem to get a personal God off the hook, either. Who created Satan? Believers have to say "God," and they refer to the myths and stories of fallen angels. But the same questions arise with angels as with humans. God knew what Satan would do and yet God created him. God is all powerful yet permits Satan to do all manner of evil things. Satan as the cause of natural evils is not a plausible, satisfactory answer to a serious objection concerning the existence of a single, all-powerful personal God.

Religions that have two equally powerful but separate deities—one all good, the other all evil—or polytheistic religions do not have this particular problem.

The book of Job provides another response from the believers.

Job, an innocent and God-fearing man, is the subject of a wager between God and Satan. God permits Satan to cause Job extreme suffering until Job cries out, demanding an explanation from God. Yahweh's answer is: How dare you question or judge God? This is an attempt to cut off debate, to stifle the use of reason and intellect, and to leave the contradictions unresolved. What kind of a God is this? Not the personal God of Collins's New Testament, who preached the message of love, mercy, and justice. Who am I to question God? I am not questioning God since I am not one of the people God chooses for conversation. I am questioning Collins and other believers about their evidence for God.

## SUMMARY STATEMENT

After consideration of these arguments for and against atheism, I am strongly inclined to accept atheism as a worldview. The arguments satisfy the burden of proof beyond any reasonable doubt. The impossibility of god, most especially a personal God, has been reduced to a point close to absolute certainty.

In the end, it is the evidence and methods of science that provide satisfying natural explanations to the universe. In fact, science generates critical irresolvable problems and contradictions for believers that actually destroy the plausibility of a personal God. Science, while limiting its scope to the explanation of the universe in natural terms, gives no support to the concept of a supernatural reality. Science discredits the supposed evidence used by believers like Collins by finding natural explanations for altruism and purported miracles. Science explains the universe in terms humans can understand. Religion tells us the universe is not explainable in human terms. It is a mystery that requires blind faith in a supernatural god beyond human understanding. For a thoughtful perspective on atheism as a rational and satisfying worldview, read *Philosophers without Gods*.[9]

Collins finds agnosticism a more tolerable, less aggressive, and logically defensible position, but regards it as a cop-out, adopted by adherents who have not really studied the evidence. He is wrong. Some agnostics have looked at a great deal of the evidence for and against the existence of God. They simply have an extremely high burden of proof. They want certainty or close to it. Some agnostics consider God's existence and nature to be beyond human understanding, at least at this point. To my mind, if God is such an unfathomable mystery, his existence has no practical relevance to humankind. Agnosticism is an acceptable position, in my view, because it is compatible with science.

## NOTES

1. F. Collins, *The Language of God* (New York: Simon & Schuster, 2006), p. 158.
2. Ibid., p. 195.
3. A. M. Nicholi, *The Question of God* (New York: Free Press, 2003).
4. Collins, *Language of God*, p. 164.
5. Ibid., p. 165.
6. G. Smith, *Atheism: The Case against God* (Amherst, NY: Prometheus Books, 1989); D. Mills, *The Atheist Universe* (Berkeley, CA: Ulysses Press, 2006); D. R. Steele, *Atheism Explained* (Chicago: Open Court, 2008).
7. C. Hedges, *War Is a Force That Gives Us Meaning* (New York: Anchor Press, 2003).
8. T. J. Wray and G. Mobley, *The Birth of Satan: Tracing the Devil's Biblical Roots* (New York: Palgrave Macmillian, 2005); E. Pagels, *The Origin of Satan* (New York: Random House, 1995).
9. L. M. Antony, *Philosophers without Gods* (New York: Oxford Press, 2007).

# *Chapter 8*

# SUPERNATURALISM (ETHICAL MONOTHEISM, SPIRITUALITY)

*Truth is great and will prevail if left to herself. She is the proper and sufficient antagonist to error, and has nothing to fear from the conflict unless by human interposition disarmed of her natural weapons, free argument and debate; errors ceasing to be dangerous when it is permitted freely to contradict them.*

—Thomas Jefferson

Collins recognizes that theistic evolution—that is, simultaneous belief in evolution and a personal creator God—is a minority position among scientists. He offers several explanations for the rarity of this belief, namely, (1) theistic scientists rarely speak out for fear of criticism from the majority of their peers; (2) *theistic evolution* is a "terrible" name (which Collins tries to correct by renaming it *biologos*); (3) the media promote conflict and frustrate the production of the harmony that theistic evolution, biologos, would produce if it were broadly adopted; and (4) many feel theistic evolution involves unacceptable compromises that do violence to science or faith or both.

## WHY IS SCIENCE OPPOSED TO CHRISTIANITY?

Let us consider Collins's explanations of why theistic evolution is a minority position in the scientific community. With respect to his first suggestion, fear of criticism by fellow scientists, it should be noted that scientists, in order to get their scientific work accepted and used, must present it to other scientists and defend it or change it in response to criticism. Professional scientists who achieve the greatest recognition are those who propose and defend new ideas. New ideas, when introduced, are almost always minority positions and are routinely attacked by scientists holding the older, more orthodox position.

If you believe that the most important element in your life, your real purpose and meaning, is found only in fellowship with Jesus, it seems to me that you would be strongly motivated out of respect for truth and out of caring for the souls of your fellow scientists to speak out and deal with the criticism. Most religions expect their believers to be evangelistic. We have freedom of speech and religion, and these rights would be protected against discrimination by federal and state law. So what would prevent you from stating your theistic belief? Collins is to some extent correct. Such scientists would be criticized for failing to behave as scientists since they have abandoned the need for logical, rational, acceptable evidence of sufficient quantity and reliability to support their religious beliefs. Their standing as scientists depends on their use of evidence and logic to support their theories. If they include supernatural or biblical evidence or arguments, they would no longer be acting as scientists.

I do not believe that Collins has lost creditability or respect among scientists when he speaks as a scientist on scientific matters, even if they do not agree with his religious arguments. I agree with Collins that scientists should overcome their reluctance to study and discuss religious beliefs and practices. Scientists can help nonscientists determine the truth or falsity of religious claims to knowledge. The fact that most scientists avoid any involvement with religion is because it is in their self-interest to do so. Too many scientists are quiet atheists because they feel they would risk loss of public support and private funding. They prefer to devote their time to scientific investigation rather than responding to the constant attacks and criticism of believers and prefer to leave the defense of atheism to the Dawkinses of the world.

Collins's second point, that the name *theistic evolution* "doesn't resonate particularly well," is not a major barrier for scientists. As Shakespeare said, "A rose by any other name would smell as sweet." I am sure you will agree that you are more concerned with the truth of a worldview than the name someone gives it. *Biologos*, Collins's suggested name, is just as irrational as *theistic evolution*. There is, of course, no disputing how language and naming can distort and influence a proposition or argument. Renaming the estate tax on multimillion-dollar estates the "Death Tax" has persuaded many unthinking average-income people (who would never have to pay it and who would benefit from its redistribution of accumulated wealth) to support its repeal against their own economic interest. But would you still support the repeal of the estate tax were it named the "Tax on Unearned, Inherited Wealth," or the "Paris Hilton Tax," or the "Rich Kids' Tax"? Of course not. These names emphasize the unfairness of the concentration of wealth in the hands of people who did nothing more to earn it than to be born with the right name. The problem with theistic evolution is its logic, not its name. Furthermore, I predict biologos will not win popular usage or change the mind of many scientists.

Collins's third point concerns the role of the media in promoting controversy and not harmony. This is a valid point in my opinion. The media outlets, with a few exceptions, are in the business of making money. Media professionals know that unusual and emotional stories provoke interest and readership. Journalists constantly run back and forth from spokespersons of opposing positions in a "he said, she said" dance that frequently confuses and seldom educates the public. Within the science community, there is no significant controversy about the basic elements of the evolutionary process. With science there is ultimately only one interpretation that corresponds to the true state of reality. Either AIDS is caused by a virus or it is not. The media generate controversy, like the evolution "controversy," by pitting religious proposals and pseudoscience against scientific facts to stimulate the audience's emotional responses.

But the media also generate attention by appealing to those like Collins who want to see reconciliation between the facts of science and the dogma of the faithful. Frequently, when a new scientific result or theory is published, the opinion will be expressed that it supports some religious belief such as "science finds evidence of great floods or the seven plagues or . . ." fill in the blanks. This implies the war between religion and science has been ended and all contradictions and conflicts between them have been or will soon be resolved. This panders to believers and misleads the public. Responsible media that sincerely desires to inform the public need to clearly separate scientific facts and their interpretations from nonscientific, especially religious, interpretations. The media also need to distinguish between majority consensus interpretations of the scientific community and controversial, minority fringe interpretations.

Finally, Collins gets to the real reason biologos is a minority position for both scientists and believers. Both groups find theistic evolution a logically incompatible, labored, and unconvincing attempt to compromise truly substantively different sets of facts and belief systems. Atheists consider it a "God of the gaps" theory, invoking the supernatural where it is not needed. Christians consider it an attack on the truth of the Bible and their concept of man as created in God's image.

But is biologos an alternative scientific theory? Evidently not. Collins, in a particularly abstruse and unintelligible passage, says, "Biologos is not intended as a scientific theory. Its truth can be tested only by the spiritual logic of the heart, the mind, and the soul."[1] He needs to decode this vague and obscure poetic message for natural scientists who regard the heart as a muscular pump with no logic, spiritual or otherwise. The mind, as the collective function of the brain, does include natural logic, but "spiritual logic" would need to be defined. He also introduces without any prior definition, discussion, or rational demonstration an entity called the soul.

I am totally bewildered as to how one would test the truth of the worldview Collins calls biologos. In order to defend biologos from the charge that it makes God an inefficient creator relying on uncontrolled chance, he argues for a deterministic universe where "from God's perspective the outcome would be entirely specified."[2] He is apparently willing to sacrifice free will and make God responsible for suffering. The universe he describes functions in accordance with natural laws just the way one would expect it to function in the absence of God.

Collins returns to the apparent conflict of biologos with his sacred texts. He resolves the discrepancies by reinterpretation of the symbolic, metaphorical text of Genesis. He concludes with a call for his readers not to turn their backs on science or faith, one to the exclusion of the other. But can one accept science, a worldview that requires only natural causes and has no need of a supernatural dimension, while also accepting the supernatural as its ultimate support? Science does not accept the supernatural, and religion depends on it. If one is true, the other is not. I agree with Collins that science and faith will be with humankind for a long time. But my prediction, based on years of polling and surveys, is that there will be a very slow but constant diminution of faith and an increase in importance and acceptance of science in future millennia, especially if you define faith as belief in a specific personal God of a specific religion.

When all is said and done, I believe Collins has not accomplished his limited objective of reconciling one possible theistic point of view, Christianity, with modern science. Nor has he achieved his broader objective that may, in fact, be unachievable: to establish that belief in a personal God is more reasonable, more plausible, more intellectually honest and satisfying; in short, more likely to be true than atheism or agnosticism.

## THE SUPERNATURAL

We know that there are natural objects, a real universe of which we are a part. I doubt that anything a theologian, philosopher, or scientist could say would cause you to deny this fact. We can only believe, speculate, and imagine a possible supernatural existence. Believers have a theory that, in addition to the natural world, there is a supernatural world. But it is "just a theory." Collins separates himself from science by his too eager and unsupported faith in the supernatural. All religions, even those that are polytheistic, deistic, pantheistic, or even abstractly spiritual like Buddhism, share belief in a supernatural realm, which by definition is beyond our senses, our comprehension, or our science. This realm has no limits on what it might contain—spirits, angels, demons, gods, souls, ghosts—who knows what? If it is beyond our knowledge (true belief based on sufficient reliable evidence), we cannot know the supernatural exists and can only imagine what it might contain. Humans are fascinated by the supernatural. It provides a feast for the imagination. A large percentage of the entertainment you enjoy (books, movies, television) is based on the use of supernatural forces or beings in order to terrorize you; aliens, vampires, zombies, and invisible malevolent spirits that obey supernatural (not natural) laws. Even though you know "it's just a movie," you can't help feeling the emotions the filmmakers are trying to generate. Maybe like me you feel a little silly when an involuntary tear appears or you find yourself gripping the arm of your chair. How willingly we suspend our disbelief and enter the narrator's weird world with its unnatural game rules, knowing normality and control will return as soon as we leave the theater.

In this real world, science has provided such good, reliable explanations for all the experiences of our lives that we need to ask, what is left that requires a supernatural explanation? This is not to say that science has the answers to all our questions about the natural world at this point in time. Reality changes with time, and knowledge about reality needs to change to reflect this. Scientific knowledge is always tentative and incomplete but it is the best and only real knowledge we have.

To establish the existence of the supernatural, you would have to show that there are phenomena or events that could never be explained by any scientist at any time in the future by the use of natural laws and causes. The fact that you cannot explain an event that you have experienced or heard about does not give you permission to jump to the conclusion that it has a supernatural cause. We have to be especially skeptical about supernatural events that are described to us. We humans love a good story that touches our curiosity or emotions. When we retell a good story, we tend to make it a little bit better. The single tear a statue sheds becomes several tears, then weeping, then grimacing or closing the eyes, and so on. I am sure that, like me, you have probably witnessed feats of magic performed by a master illusionist. We cannot come up with any natural explanation. In fact, the illusion seems to be impossible without a violation of natural laws. But we all know that there is a very good natural explanation for what we see.

If the supernatural is so very different, so unnatural, so separate from the natural world of which we are a part, then we have to ask, can it ever have any effect on our natural world? Is it so different, so incomprehensible as to be undetectable? It would then be irrelevant to us. There is no point in arguing about something that has no impact on natural events. On the other hand, believers in the supernatural may claim that it does interact with our natural world either continuously or intermittently. If this is true, then scientists studying the natural world should find evidence of a supernatural effect. A physician, Yonatan Fishman, has asked the question, "Can Science Test Supernatural Worldviews?" and responded in an article with this title in the online journal *Science and Education*.[3] I agree with his affirmative response. We reject the contention of Stephen Jay Gould and some scientific organizations that the supernatural is beyond the reach of scientific investigation. Fishman lists "three ways to evaluate the truth of a claim, (1) by consideration of a prior probability of the claim being true, (2) by 'looking and seeing' (i.e., by considering the evidence for or against a claim), and (3) by consideration of alternative explanations of the evidence." He uses Bayesian probability to make the point that when either the probability of a supernatural is very low or the evidence that should be found if the supernatural explanation were true is not found, then the absence of evidence is indeed evidence for the absence of the supernatural.

We should consider the fact that all the evidence collected about the universe is consistent with a naturalistic worldview and the fact that there is an absence of evidence supporting the supernatural is supportive of the natural and not the supernatural worldview. Fishman, after an excellent review of the question, concludes, "The best explanation for why there has been so far no convincing, independently verifiable evidence for supernatural phenomena, despite honest and methodologically sound attempts to verify them, is that these phenomena probably do not exist."

## IS THERE EVIDENCE FOR THE SUPERNATURAL?

Let us consider the relationship of the natural to the supernatural. No one but a complete skeptic can deny the existence of a natural universe. Even if it is denied as a matter of belief, you have to acknowledge its existence and laws in practice or you would soon perish. The natural universe contains all the things we experience and perceive directly, for example, stones, or by perception of their effects, for example, gravity. Believers in a personal God all believe in a supernatural reality that exists separate from natural reality, because God is a supernatural being. As a supernatural being, God could assume a number of different relationships to the natural universe we inhabit. God could exist totally separate from the natural world and never interact to alter natural events. This belief results in a deity that is irrelevant to any human concerns or behavior. Science could have no way of observing, proving, or disproving this kind of god. This is not a god that inspires or desires worship or fellowship. This concept of an impersonal god who created the natural universe and sustains it by means of the natural law is called *deism*. Again, this is not a god that especially cares about or interacts with humans. Scientists could not distinguish between natural and supernatural because they would be studying only one unvarying set of divinely decreed and sustained natural laws.

On the other hand, Collins's personal, supernatural God would have to make his presence felt in the natural universe by interacting with his creatures, namely, mankind and its natural laws. But then you would think there must be some changes in the uniform action of natural objects or processes that are potentially observable by humans and therefore a potential subject for scientific investigation.

As long as a natural explanation of an event or phenomenon is possible, it is more probable, more rational, more in keeping with our experience than a supernatural one. Even Collins would agree that, for example, most spontaneous cures of natural diseases and disorders are due to natural causes, and only a few would he claim as miracles. Collins agrees that unexplained events are not unexplainable events, and believers should not too quickly propose that God fills the gap in knowledge and explains the event. Any unexplained event might, in the future, be explained using natural laws. It appears that the only proof of supernatural intervention in the affairs of the universe would be an event that was totally unexplainable and violated all natural laws or explanations.

When considering how our natural universe might interact or be affected by the supernatural realm, we need to consider whether there is a different set of supernatural laws that order the spiritual world, however different from natural laws they may be, or whether the supernatural world is chaotic, disorderly, and incomprehensible. Most believers suppose that there are uniform spiritual laws that control events in the supernatural realm. An unpredictable, unrelated, unreliable spirit world, not influenced by humans or spirits or supernatural events, would eliminate any knowledge of souls, miracles, or spiritual rewards and punishments. Everything in the supernatural becomes arbitrary and unknowable from a human perspective.

Now, if there are some laws, some regularities in the way the supernatural world operates, it should be possible to learn what they are. The supernatural laws could then be integrated into our knowledge of natural laws and expand the area for scientific investigation and explanation. The supernatural would interact with the natural laws to produce effects that science could study. Evidence could be collected that supported the theory of a supernatural realm with its own laws.

To deny that the supernatural leaves evidence of its existence is to deny that belief in the supernatural is intelligible or rational and leaves as the only alternative blind faith unsupported by any evidence as the basis for belief in the supernatural. Alleged scientific support for the claims of Christianity is more to the point. In Christianity, the supernatural power of God is expressed in the universe by such things as:

1. Christian miracles, events that are beyond any reasonable scientific explanation and violate the natural laws so that the only possible explanation is a supernatural one. (Miracles and the evidence Collins uses to support them was discussed in chapter 3.)
2. Response to Christian prayer by successfully obtaining changes in natural events. (Possibly better response to Christian prayer over Islamic or Jewish or Mormon prayer.)
3. Communication with the dead using supernatural means or having experiences that seem to transcend natural death.
4. Human demonstration of supernatural powers, such as psychic powers of various sorts, for example, extrasensory perception or mystical, supernatural experiences, which have no natural explanation and defy natural laws.
5. Supernatural visions or apparitions like Our Lady of Fatima, Our Lady of Lourdes, Our Lady of Guadalupe.

## MIRACLES AS SUPERNATURAL EVENTS

Collins attempts a defense of his belief in miracles by appealing to probability, specifically Bayesian probability. The basic purpose of Bayesian probability is to see how the known or assumed probability of one event, given one set of circumstances, will be affected by changing some of the circumstances. We used Bayesian probability in the California prenatal screening program. First, you determine the probability of a woman giving birth to a child affected by Down's syndrome, which is based on age. The older the woman, the higher the risk. So, for all thirty-five-year-old women, on average 1 in 365 births will be affected.

But there are other things beyond age that increase or decrease the risk. The concentration in the blood of certain chemicals called *markers* also can increase or decrease the chance of a Down's birth. We take the previous known age-related risk, called the *a priori*, or original risk, and modify it to make a better estimate of the risk for each individual thirty-five-year-old so that within the whole group of thirty-five-year-olds some will have lower risks and some will have higher risks than the average for the whole group. This gives us the most accurate estimate of the risk of Down's syndrome for this particular woman. This is called the *a posteriori*, or adjusted risk.

Collins uses the "apparently miraculous" cure of a cancer "known to be fatal in nearly every instance" as an example of a miracle. Since scientists have not had the opportunity to study every single case of this cancer, the chance of a spontaneous but naturally caused cure cannot be known independently. The a priori chance of a spontaneous cure is very low, but not zero. But what are the chances the cure was due to supernatural intervention by God?

If you don't believe in God or the supernatural, then Collins correctly identifies the a priori probability at zero, or totally impossible. Collins says you have to allow some probability, no matter how small, and therefore miracles, however rare, are theoretically possible. The same can be said about the existence of a herd of Danish-speaking feathered cats somewhere in Antarctica. Possible, but definitely not proven or even very plausible. The probability of a supernatural cause for an event is always lower than the probability that the event was incompletely studied or inaccurately reported. For example, the probability of misdiagnosis is higher than the probability that God did it. The same can be said of the probability that this was a form of cancer with a higher spontaneous cure rate, or that the cure was only a temporary remission and the patient later died, and so forth. As a physician, I am aware of the wide variety in response to injury and illness but have never seen or heard of anything that required a supernatural explanation.

What is the probability you would survive a fall of three feet? Very high. Probably not a miracle. What is the probability that you would survive a fall of 33,000 feet? Vesna Vulovi, a flight attendant, survived such a fall strapped in a seat in the tail section of a plane blown up by a terrorist bomb. She was the sole survivor and, in spite of skull fractures, two broken legs, and three broken vertebrae in her spine, one of which left her temporarily paralyzed from the waist down, she recovered and resumed working at the airline.[4] A miracle? Science cannot provide a detailed analysis of the physical forces that combined to result in this highly improbable outcome, but that does not mean that it is a violation of the natural laws. If she had been saved by supernatural intervention, I would have expected a clear violation of natural law, such as she was unbuckled and walked away from the wreckage unharmed.

It is also curious that modern miracles are always attributed to positive events. There are no reports that everyone on a plane lived but one person miraculously died. No one is miraculously killed by a lightning strike. Have you ever heard anyone say it's a miracle that a tornado hit his or her house and missed the one next to it?

Collins refers to miracles in the Bible as evidence for a personal God. The Bible reports that Jesus and his disciples, who were not divine, performed miracles. If Jesus enabled his disciples to perform miracles, perhaps someone else did the same for him. So, performance of miracles is not evidence of Jesus' divinity. Ancient texts are full of stories of miracle workers. It was an age of miracles. Moreover, most miracles described in the Bible are about casting out demons, healing sick, lame, paralyzed, mentally ill, deaf, mute, or blind people. Modern-day faith healers, many of whom have large and gullible followings, make exactly the same claims. The effectiveness of strong belief in a healer or a procedure can mobilize natural curative mechanisms of the body. This is called the *placebo effect*. If Jesus could miraculously heal the sick, this raises the question, why didn't he heal everybody of everything? This would be consistent with Jesus' nature of universal love and would have certainly convinced Jews and Gentiles in his time that he was divine and his teachings were true.

In defense of miracles, Collins speaks only in general terms, but, as so often happens, the devil is in the details. Consider the miracles described in Acts 5:1–10. The apostles were raising money to support their new Christian sect. A Christian follower named Ananias sold a piece of property but did not contribute the total proceeds of the sale to the apostles. Peter accuses him of keeping part of the money, and when Ananias denies this he is immediately, miraculously killed. His wife, Sapphira, appears three hours later, and when asked by Peter, she also lies about the price collected. She is immediately, miraculously killed. The human author of Acts was clearly writing to propagandize for his version of Christianity. His purpose in these verses was successful in that "great fear came upon all the church, and upon as many as hear these things." Of course, there is the distinct possibility that this does not report an actual historical event or that it changes the circumstances. Perhaps the time of death was several days later, or both were victims of food poisoning. But, if true, how do you explain God's purpose in directly intervening by imposing the death penalty for such believers for a minor or mercenary offense? If the offense was serious—lying to God—why weren't all others who lied to God also instantly killed?

Another miracle is Paul's blinding of the sorcerer Elymas for opposing his teaching (Acts 13:8–12). These kinds of miracles do not support Collins's and Lewis's contention that miracles are events of special significance designed to highlight divine purposes or teach spiritual lessons.

As part of my intellectual journey from Roman Catholicism to atheism, I looked at miracles as a way to buttress my faith. While in medical school I purchased a small volume detailing evidence for medical miracles accepted by the Vatican. *Modern Miraculous Cures* was an English translation of the work of two religious Catholic French physicians who, in their preliminary declaration, acknowledged their submission to the judgment of the Holy See with respect to "every interpretation or theory contained in their work."[5] Much of the book concerns documentation collected at Lourdes. Catholics believe that after the Blessed Virgin appeared to peasant children at the baths of Lourdes, the springs there then produced miraculous cures. Nothing in the various cases documented could rule out natural reasons for the cures. Even when a cure was unusually rapid or otherwise unexpected, nothing, if completely and objectively investigated, would support a clearly supernatural exception to the natural law. Several questions went unexplained. Why were a few cured but most not? Why, after receiving a cure, were the afflicted not converted to Catholicism? Why did some of the sufferers remain cured while others had a recurrence of disease? Finally, what divine lesson do these cures impart?

Just because these doctors writing in the 1940s could not fully describe the causes of the changes that the pilgrims experienced does not require us to attribute the changes to the uniquely Catholic God of the gaps. If Collins and other Christian believers accept biblical reports of miracles and are convinced that the miraculous cures at Christian sites are true, don't they also have to accept the possibility that the miracles of Islam and Hindu gurus are also genuine miracles unexplainable by natural means? What do these non-Christian miracles tell us about God's message and the seal of authority of one true religion?

## PRAYER

I am only discussing intercessory prayer that asks God to intervene in the natural world. Prayers that ask for supernatural favors like forgiveness of your sins or blessings for your dead relatives cannot be tested. I am not concerned with prayers that thank or praise God or just reinforce one's commitment to, or are meditations on, God. Petitionary prayers, where persons pray for relief of pain or some such physical benefit for themselves, are difficult to test because science has demonstrated the placebo effect. If a person strongly believes that some action will have an effect on the body, this belief will trigger a physiological reaction with a real effect. Intercessory prayer made for the benefit of others is the only kind that could provide evidence of the supernatural. Believers will point to answered prayers as evidence of a supernatural, personal God. However, they fail to appreciate that the overwhelming majority of prayers go unanswered. These believers are guilty of the optimistic human fallacy of counting only the hits when something happens after prayer and not the much more numerous misses.

People pray for opposite and incompatible outcomes. The farmer asks God for snow to provide needed irrigation water, while the rancher prays that snow won't fall because it would kill his mountain-bound herds. It is guaranteed that one but not the other will have his prayer answered. During the Hundred Years' War, history records the spectacle of pious French and Spanish Catholics praying for the slaughter of English Protestants and English Protestants praying to the same Jesus Christ for the destruction of the heretical Catholics.

If prayer is really effective, why do we not pray for true miracles? For example, why doesn't a person with limited vision pray for the ability to accurately see anything at infinite distance through any intervening material, instead of just praying for the restoration of normal vision? The former would be an indisputable demonstration of the supernatural, while the latter could always be explained as a unique but natural event. We pray for naturally possible things like the ability to walk again. We do not pray for the amputated leg to grow back. This is proof that we don't expect a real miracle. For hundreds of years millions of people's earnest prayers for peace have not prevented wars that continue to occur with increasing brutality and casualties. Prayers have not ended cancer or prevented AIDS. I am also curious as to why we pray for the dead. If souls exist and are judged worthy of heaven or deserving of hell, the judgment is final. Praying for a loved one will not free her from eternal punishment or add to her eternal happiness. Maybe prayer has more to do with its effect on humans than its effect on God.

There have been several studies of the efficacy of prayer, most of them poorly designed and controlled. Better studies by reputable scientists have not found any evidence for the efficacy of prayer. An example is the large study of the effect of prayer on mortality and complications following cardiac bypass surgery.[6] When the 604 patients who received prayer were compared to the 597 patients who did not, the mortality was the same in both groups. The complication rate was higher in the group receiving prayer than in the control group. This $2.4 million study was funded by the Templeton Foundation in a failed effort to get scientific evidence for the supernatural.

When positive results are reported, they are insubstantial, such as reduction in anxiety; statistically insignificant; or have clear natural explanations. None of these studies done by believing scientists have reported the dramatic results than would be expected if prayer were powerfully effective. The only exception to this statement was a controversial evaluation of prayer conducted in Korea. This study, published in 2001, reported there was 100 percent successful treatment of infertile women when Christian prayer by Americans, Canadians, and Australians was added to the treatment.[7] Of the 100 women who received prayer, all became pregnant as compared with only 26 of 100 women who did not receive prayer. The study was repudiated by one of the coauthors, Mr. Daniel Wirth. He is a con man who was later convicted of mail fraud. Another coauthor, Dr. Rogerio Lobo, denied participation in the research and formally requested his name be removed from the publication. The last remaining author, Dr. Kwang Yul Cha, defended the study and even filed suit for defamation against one of the study's major critics, Dr. Bruce Flamm. The courts dismissed the suit as without merit. This study, which is still cited as evidence by believers, has not been confirmed by any independent investigation and is widely regarded as bogus.

Incidentally, the designer of human reproduction could hardly be characterized as good or intelligent. Only 25 percent of the time will unprotected intercourse result in a pregnancy. Between 40 and 65 percent of pregnancies end in miscarriage. About 40 percent of the women develop complications, 15 percent of which result in chronic problems. The World Health Organization reports that worldwide over half a million women, 1,500 every day, die from pregnancy-related causes.[8] WHO reported that 3.3 million births end stillborn and 4 million newborns die in the first twenty-eight days of life. Premature birth is a big contributor to this figure. Currently they estimate 40 percent of women worldwide receive prenatal care. The toll of natural pregnancy, as designed by God, in the absence of modern prenatal care would be even higher. About 3 percent of live births have serious birth defects.

This does not impress me as the work of an intelligent designer. It is grossly inefficient, and if you had the power to redesign the process, I am sure you could do a better job. If the process had a designer, I would not consider the designer to be a good person in view of all the misery attached to the cited statistics. Why cause all this worry, pain, and grief to your most cherished creatures? I have attended hundreds of deliveries and know firsthand what a powerful, life-affirming, and joyful experience this can be when the process results in a beautiful baby. I am also familiar with the numerous problems associated with its "design." While the world contains awe-inspiring and beautiful mountains, rivers, lakes, beaches, and forests and marvelous, well-adapted life-forms, it also contains ugliness and failures to adapt associated with poor design. Human reproduction is only one example of the imperfection of the universe's design and operation.

Scientific studies of prayer have been challenged by some believers based on their belief that God does not perform just to satisfy the curiosity of scientists. God resents the fact that God's power is being tested. Why would God not grant prayer for the good of others? What better way for God to make known God's existence and to encourage humans to take God's Gospels seriously? Why would

God not look upon intervening in response to tests of prayer as an opportunity to win over doubters? The Gospel of John reports Jesus performing miracles in response to requests for signs to convince enemies or doubters. Did God resent Thomas, who doubted Jesus' Resurrection? Doesn't God want to have fellowship with honest doubters and save their souls from eternal suffering in hell?

This reluctance of God to cooperate with unbelievers looking for evidence of God's existence has, in fact, been used as an argument against the existence of God. Based on Collins's claim that God desires fellowship with humans, the argument can be stated as follows: If there is a loving, omnipotent God who desires fellowship with all humans, wouldn't God provide an effective means for all humans to recognize his desire for fellowship? Now, it's true that the majority of the world does not believe in the Christian message, so despite many opportunities and the availability of methods that do not involve coercion, God has not provided effective means to all humans to recognize his desire. Therefore, either such a God does not exist or he is not a personal God desiring fellowship. Why does the Christian God allow atheists, agnostics, Muslims, and truth-seekers to look in vain for clear signs validating the important message he has for humankind? Why is it so hard to hold on to logic and the truth of our experience, our reason, and our science, and to believe in the resurrected Jesus or any other personal God?

I would like to outline an experiment to test the efficacy of prayer. The subjects must not know whether they are being prayed for. The subjects must be atheists, agnostic, or not believe in a personal God, but be unaware that this is a criterion for inclusion in study. This ensures they will not pray for themselves. The experiment should include a randomly assigned, experimental group that receives prayers and a control group that does not. The groups need to be of sufficient size to have statistically significant results. The groups should all be matched for sex, age, race, income level, and health status. The effect of the experiment should be clearly good both physically and ethically. The people praying should all be from the same religious sect, for example, Baptists, Sunni Muslim, and so on, and should say a standard prayer for the unknown subjects using code names. The standard prayer should include a request for the sake of their subject's soul, for God to provide a sign to all unbelievers of his power and goodness that could not be explained by natural laws, for example, that for a brief time they would be able to see and read verses from the Bible sealed in a lead box.

The objective criteria for success should be defined before the experiment begins and should be unknown to the investigators conducting the actual experiment. The team of investigators should not know which subjects are receiving prayers and which are not. The data collectors should be professionals who do and do not believe in the efficacy of prayer and should include some atheists or agnostics or spiritualists who do not believe in a personal God. The investigators should be blind during data analysis. The identity of the groups should be revealed only in the last stage of collating results.

Another example might be praying for a group of animals on which a vaccine for HIV or malaria was being tested. Infections, length of temperature elevation, survival, and number of parasites per milliliter of blood on specific days after injection could be used as objective measures of effectiveness. If the vaccine were effective, it would be a major benefit for humans. As a strict demonstration of God's response to prayer, and as a sign to nonbelievers, you could use a dose of virus or parasites that would kill 50 percent of animals.

Believers sometimes argue that God does not give signs or perform on cue. I suppose these believers do not use petitionary or intercessory prayer, but it begs the question how reasonable is it to believe in a God who wants and demands belief without clear signs of any kind? If Jesus or Allah or Yahweh or Zeus does not perform on cue, what motivates you to pray to them? To believe in the absence of consistent hard evidence opens up belief in infinity of weird and impossible things.

## PARANORMAL CLAIMS

There are a wide variety of claims of unusual events or mental experiences that could be classified as paranormal. These include paranormal powers such as clairvoyance, telepathy, extrasensory perception, psychokinesis, precognition, communication with the dead, and so on. This also includes paranormal experiences like out-of-body, near-death, and mystical experiences. Most claims are isolated reports that are never investigated. Because of their intriguing nature, groups have been formed to study them. The Society for Psychical Research began studies in 1882 and major universities have had organized paranormal research groups. The results have been minimal and the methods of data collection and interpretation have been widely criticized. Parapsychology has not been scientifically accepted on the basis of currently available evidence.[9] Frequently, knowing the intense interest in mystery, the unusual, and the supernatural, the media report such events as if they happened exactly the way they are described by the "psychic" or the witnesses. If you read it in the newspaper or see it on TV, it "must be true" and requires an explanation. The feats performed by alleged psychics, like identifying cards, changing the results of a random number machine, and bending spoons, are unimpressive accomplishments. This hardly seems the sort of thing that requires divine intervention. If they occur, they only affect a few individuals and are very uncommon events. Rarely do they have any religious significance. They do not provide evidence of the existence of the supernatural beyond a reasonable doubt.

One of the supposed expressions of the supernatural is the phenomenon of near-death experience (NDE). People whose vital functions have temporarily stopped before they are resuscitated (not resurrected) report having one or more of the following memories: separation from their body, bright lights, passing through a tunnel, seeing deceased loved ones, and having memories of their past activated. The frequency of NDE is in dispute, but the best estimate from medical data is it occurs in 10 percent of survivors.[10] It is interesting that very few reports of

NDE come from areas where life-saving technology is not available and used, raising the possibility that these are natural phenomena. Some NDE are actually false memories of an event that did not occur.[11] There is no way to predict who will have an NDE. They can happen to anyone who has a brain. These are real experiences and have to be satisfactorily explained and not just explained away.

There are obvious problems in investigating NDEs since they are unpredictable and cannot be exactly duplicated by risking death. First, the experiences are private, subjective experiences involving primarily the emotional and visual centers in the brain. After the event, the thinking and speaking modules try to reconstruct the experience into some semilogical order. The experiences are ineffable—they are difficult to put into words. We all have such experiences like falling in love, the sensation of orgasm, the thrill of victory, or the agony of defeat. This makes it difficult to define and quantify the emotional dimensions of the experience.

Second, a person is really dead some time after there are no signs of life as measured by detectable electrical activity of the heart, called an electrocardiogram or EKG, and the failure to detect electrical activity in the brain, called an electroencephalogram or EEG. The minimal amount of brain function and blood needed to sustain life may not be measurable by current medical equipment, but examples of survival after instruments show no activity are not infrequent. Just when in the dying process the NDE occurs is not known, but it may be before or after the EEG is flat. NDE are altered states of consciousness.

One of the widely cited investigations of NDE was done in the Netherlands on 62 survivors of cardiac arrest.[12] They listed ten elements associated with NDE and classified survivors based on how many elements were present. Thirty-three percent had minimal elements, 29 percent had moderate, 27 percent had deep, and 9.6 percent had very deep experiences. This seems to indicate a progression of the depth of the experience as stress and loss of normal function increase during the dying process. Do these NDEs require a supernatural explanation?

Doctors and scientists who have investigated NDEs have provided evidence that they are natural phenomena.[13] An essential cause of NDE is severe physical or psychological stress. Lack of oxygen to the brain, which might differ in different areas of the brain, also plays a role in most NDEs. This would account for the variability of elements reported. A variety of observations provide evidence that the best explanation of NDE is as a natural phenomenon. Elements of NDE can be produced by electrical stimulation of the temporal lobe of the brain.[14] Administration of drugs such as LSD and ketamine produces a very similar experience.[15] Fighter pilots who black out in steep dives report elements of NDE. Victims of brain injury report some elements of NDE.[16]

These naturally induced experiences closely reproduce almost all of the elements of an NDE, but without the added stress of the simulated NDE situation, cannot be expected to be identical. Those who would use NDE as an example of a supernatural phenomenon claim the brain changes are only a means to unlock access to the supernatural. They have to account for why only a minority of people who survive a life-threatening situation have an NDE. What distinguishes the select few? Again believers are arguing that if science cannot completely explain NDE, then by default it must be supernatural. They also have to explain why the visions are culturally conditioned. Muslims see Muhammad and Christians see Christ. Part of the feeling that these are supernatural events comes from the fact that a great many of the survivors have markedly changed their lives after the NDE. It is not surprising that almost dying would change your approach to life. However, while in most cases the changes might be described as positive, there are also some that have negative changes. One investigator who did a ten-year follow-up study of NDE found that NDE "may include long-term depression, broken relationships, disrupted careers, feelings of alienation, an inability to function in the world, long years of struggling with the keen sense of an altered reality."[17]

## MYSTICISM

Other evidence suggested for the supernatural is the experience of mysticism, the personal experience of the divine, for example, the voice or presence of Jesus. These experiences are plausibly explained as mental aberrations in specific brain activity, which can produce feelings of transcendence, joy, and auditory and visual hallucination. The perceptions, which are quite real to the person experiencing them, are not obviously reflected in changes in the external world. Most readers will not personally have experienced the power of the mind to distort external reality. Few will have directly experienced something supernatural. The ability of the mind to experience this transcendental feeling by natural means was discussed in chapter 4.

As I mentioned, UCLA, where I received my medical training, had a strong emphasis on psychiatry. We visited mental hospitals every year and I had firsthand opportunities to hear patients describe their delusions about the real world. But my most unforgettable episode was in the regular hospital. One night I was called to assist in an emergency situation. A woman who had had a radical mastectomy with rubber drains sticking out from between her ribs was standing on her bed fending off a resident and a couple of nurses while she desperately ripped at her chest wound. She was screaming that there were rats inside her chest. We subdued and sedated her and repaired the damage, but the point is she was behaving exactly as you or I would behave if we had rats in our chests or were convinced that they were in there. To her the rats were very real. This is why people swear they see statues crying or why Teresa of Avila heard Jesus talking to her. The mind's ability to produce its own internal reality at variance with the objective external reality shared by the rest of the world is not to be dismissed with a label, but it needs to be understood as a natural, not a supernatural, phenomena.

William James (1842–1910), the Harvard philosopher and psychologist, in his study of mysticism outlined in *The Varieties of Religious Experience* noted that mystical experience typically (1) is brief and temporary, (2) comes over a person as an uncontrollable feeling, (3) is an altered state of consciousness, and (4) is impossible or very difficult for the mystic to describe in words.[18] These mental states are frequently produced by procedures that are stressful on the brain, such as fasting, drugs, sleeplessness, exhausting physical activity, and sensory deprivation. There is no reason to believe that mystical experiences conform the mind to perceive any real existing phenomena or being. These personal experiences have a powerful emotional impact, and mystics find them compelling and "true." No facts or evidence or contradictions will convince them they have not had a direct experience of the supernatural. But they are not of much help to atheists or believers who have not had such an experience.

The mystic does not acquire knowledge, as the term is defined, except of a subjective and emotional experience. Mysticism is rare and has a natural explanation. It is not evidence of the supernatural. Non-Christian mystics have demonstrated that the brain can exert control over bodily functions, slowing their heart and breathing rates and even lowering body temperature. They stress the limits of the natural, but they do not do anything supernatural. If the heart and breathing were stopped for a long period and then resumed with no ill effects under controlled conditions, this might encourage acceptance or investigation of the supernatural theory.

## SUPERNATURAL VISIONS

Supernatural visions, such as the reported apparitions of the Virgin Mary, saints, or Jesus, have been shown to be hoaxes or misinterpretations of natural phenomena or the hallucinations of pious peasants. The unpredictability of these events has prevented any systematic scientific investigation. Again, we find precious little scientifically acceptable evidence of the supernatural. If by definition science is founded on the examination and explanation of the universe by natural means and arrives only at completely natural results, what role could science play in commenting on the supernatural? There are two possible areas of interaction. First, scientists could look for phenomena or events for which no possible natural explanation could be imagined. Such phenomena, while falling short of proof, would be a point in favor of the theory of the supernatural. I am unaware of any such phenomena being documented or witnessed.

Second, when believers, as they frequently do, claim that they have scientific evidence for the supernatural, science can verify or falsify the alleged evidence. As far as I know all such "evidence" presented to make their claim seem rational and intellectually responsible has been falsified when thoroughly investigated. Religions that tie themselves to science run the risk that as science changes and as gaps in knowledge of the natural world are filled, the beliefs and doctrines of the religions must undergo change and the supernatural element is diminished and weakened.

Generally, most believers and scientists agree that science cannot prove the existence of the supernatural, but for a moment let us look at the arguments of those who dissent. The popular press, recognizing the deep-seated interest in the topic, from time to time assaults us with headlines about science and religion. In the last ten years, *Newsweek* magazine has had 906 articles mentioning God; *Time*, 2,467; *U.S. News & World Report*, 4,440; Google has 71,300,000 entries on God; and Yahoo! 76,700,000. Headlines like "Big Bang Consistent with Bible" overlook logical inconsistencies and neglect to mention many creation myths of other religions for which it is also consistent. The "consistence" requires some implausible interpretation of the science or the scripture. "Science Finds God," proclaimed *Newsweek*, July 20, 1998. This article is not a scientific proof of God or even a plausible argument for God, but an admission that science cannot and will not be able to explain everything in our lifetime. Some people, when they run out of natural explanations, or out of scientific ignorance, immediately jump to a supernatural explanation rather than develop theories and experiments that would explore the phenomena at a deeper level of nature.

As a practical matter, the supernatural does not seem to affect the universe or humankind very much. We mostly continue to act in narrow, self-centered circles, sleeping, eating, and carrying on routine activities with success and disappointment exactly as if there were no God and no supernatural reality. The methods of science produce highly probable, practically useful explanations of what the universe is made of and how it works. Scientists use human perception and intelligence and do not share Collins's need to imagine a supernatural reality to live our lives in this world.

## NOTES

1. F. Collins, *The Language of God* (New York: Simon & Schuster, 2006), p. 204.
2. Ibid., p. 205.
3. Y. I. Fishman, "Can Science Test Supernatural Worldviews?" *Science and Education*, http://www.naturalism.org/can%20Science%20Test.
4. http://en.wikipedia.org/wiki/Vesna_Vulovi.
5. F. Leuret and H. Bon, *Modern Miraculous Cures* (New York: Farrar, Straus and Cudahy, 1957).
6. H. Benson et al., "Study of Therapeutic Effects of Intercessory Prayer (STEP) in Cardiac Bypass Patients: A Multicenter Randomized Trial of Uncertainty and Certainty of Receiving Intercessory Prayer," *American Heart Journal* 151 (April 2006): 934–42.
7. K. Cha, D. Wirth, and R. Lobo, "Does Prayer Influence the Success of In Vitro Fertilization?" *Journal of Reproductive Medicine* 46 (2001): 781–87.
8. World Health Organization Facts and Figures from the World Health Report, 2005.
9. B. Farha, ed., *Paranormal Claims: A Critical Analysis* (Lanham, MD: University Press of America, 2007).
10. M. Crislip, "Near Death Experiences and the Medical Literature," *Skeptic* 14 (2008): 14–15; P. van Lommel, V. Meyers, and I. Elfferich, "Near Death Experience in Survivors of Cardiac Arrest: A Prospective Study in the Netherlands," *Lancet* 358 (December 2001): 2039–45.

11. C. C. French, "Dying to Know the Truth: Visions of a Dying Brain or False Memories?" *Lancet* 358 (December 2001): 2010–11.

12. Van Lommel et al., "Near Death Experiences in Survivors of Cardiac Arrest."

13. B. Greyson, "Biological Aspects of Near Death Experience," *Perspectives in Biology and Medicine* 42 (1998): 15–32.

14. M. A. Persinger, "Near-Death Experiences: Determining the Neuroanatomical Pathways by Experiential Patterns and Stimulation in Experimental Settings," in *Healing beyond Suffering or Death*, ed. L. Bessette (Beauport, QC: MNH, 1994).

15. K. Jansen, "Using Ketamine to Induce the Near-Death Experience: Mechanism of Action and Therapeutic Potential," *Yearbook of Ethnomedicine and the Study of Consciousness* 4 (1995): 55–81.

16. J. Bolte-Raylor, *My Stroke of Insight* (New York: Viking Penguin Books, 2008).

17. N. E. Bush, "Is Ten Years a Review?" *Journal of Near-Death Studies* 10 (1991): 5–9.

18. W. James, *The Varieties of Religious Experience* (New York: Touchstone, 1997).

## *Chapter 9*

# A PERSONAL GOD?

*It was, of course, a lie what you read about my religious convictions, a lie which is systematically being repeated. I do not believe in a personal god, and I have never denied this but have expressed it clearly. If something is in me which can be called religious then it is the unbounded admiration for the structure of the world so far as our science can reveal it.*

—Albert Einstein

One of the events that started Collins on the path from atheism to theism and then to Christianity was his decision to read C. S. Lewis's *Mere Christianity*, and the appeal of a universal law of right and wrong. My journey in the opposite direction began with acceptance of the Moral Law and the existence of sin as a fact of life. For me it was important to know and love Jesus as a person. I, of course, wanted to do what the loving Jesus Christ instructed me to do. But in my interactions with others in the Catholic community, including priests and nuns, it seemed to me that many followed the Moral Law either to avoid the threatened punishment of hell or to gain the eternal reward of heaven.

When I was still a believer, I felt that being moral was a personal expression of gratitude for the ordeal the person Jesus willingly suffered during his Crucifixion to cleanse me of original sin. My vision of the Passion and the Crucifixion was very much like Mel Gibson's film *The Passion of the Christ*, but when I was young I thought it was a special punishment uniquely applied to Jesus. In fact, in Roman times, many thousands of people were crucified as Jesus was, and none of them willingly, including Jesus. Virtue seemed to me to be its own reward and I spent little time worrying about hell or anticipating heaven.

As I grew up, I noticed little difference between the moral behavior of Christians or Jews or Hindus or Muslims or Taoists. Reciprocal altruism (you scratch my back and I'll scratch yours) was not uncommon, but sinners (you scratch my back but I don't scratch yours) were much more common than saints (I'll scratch your back even if you don't scratch mine). The interpretation of the Moral Law as proclaimed by Jesus was apparently not uniform, and continuous observance was extremely difficult.

## WHAT IS A PERSON?

Since *The Language of God* is concerned with defending the rationality and plausibility of a personal God, Jesus, we need to think about what is meant by "personal God." What do we mean when we use the word *person*? Well, obviously inanimate objects are not persons. So, being alive is essential to the concept. We don't use the word to describe life-forms that lack consciousness, like bacteria, worms, coral, and so on. When we apply the term to higher animals, we associate it with development of increasing emotional responsiveness. The most basic emotions are fear and anger. Fear was first represented in living organisms by instinctual behavior. Simple organisms moved away from light or heat or motion. This has obvious survival value but is not sufficient to make the organism a person.

It is debatable whether a snake or a lizard has a personality, although they have interesting but pretty limited behavior. One of my sons got a rosy boa constrictor as a pet. If you moved your hand quickly toward it, it would strike out because of fear, not anger. If you picked it up carefully, the snake was not concerned and could be handled. If you put a mouse in the terrarium and the snake was not hungry, he ignored it. Even on the occasion when the mouse bit the snake, it responded in momentary anger or pulled away in fear but did not aggressively pursue the mouse. For a snake every day is Wednesday, a cycle of sleeping, eating, hiding, and hunting. The snake did not love its owners and we did not love it. Snakes do not love one another or their young. A snake is a conscious, sentient animal with basic emotions, but it is not a person.

When we move on to mammals, we begin to use the word *personality*. We had pet rabbits, hamsters, and guinea pigs that had minimal personalities, but our dogs and cats had definite personalities. The beginnings of the emotional attachment called love appear as an element of their personalities. Dogs display a full range of emotions. We can communicate with them and they with us. But are you willing to call a dog a person in the same sense as a human person? Even when considering humanized, trained chimps and gorillas, there is a clear distinction.

When we refer to a human being as a person, we imply that the individual is conscious of his or herself, lives with a past and a future, and is capable of interacting with other humans and the environment in a meaningful, rational, and empathetic way. A human person is capable of assigning values and meaning to people, objects, and events in his or her experience. This is what distinguishes human beings from our animal neighbors. Humans have minds capable of reasoning, producing order, and acting with conscious intent and purpose. We are the only animal that is aware of our own mortality and the only animal that commits suicide.

You are a certain kind of person with a unique personality because your brain functions in a unique way that is characteristic of you. Your character is part of your personality. Your brain function, your personality, can be changed by dramatic events such as loss of sight or hearing; survival of a life-threatening event; sexual abuse by a trusted adult; being imprisoned, isolated, and tortured; or witnessing or participating in the brutality of war.

Anyone whose loved one's brain has suffered the ravages of Alzheimer's disease or severe brain injury can tell you that the loved one's personality has dramatically changed. If we want to use the word *person*, we need to agree that this is essentially describing a function of the human brain.

It is worthwhile to try to clarify the concept of personhood by considering the difference between a "human being"—an existing, wholly or partially complete human organism—and a human person. A thing is human if it possesses an identifiable human genome. For example, a single-cell zygote, formed by the fusion of a sperm and an ovum at conception, is a human being. It has no brain so it cannot be classified as a human person, notwithstanding that at some future time it may become a person. The zygote develops for fourteen days without developing nerves or sensation. This is the preembryo stage. The preembryo is a human being since it is an organized collection of human cells, but it is not a person. Preembryos have no brains, no consciousness, no self-awareness, no emotions, and cannot be considered persons. Personhood is gradually acquired as the embryo implants in the uterus and begins to develop into a fetus.

The point at which we can call a fetus a person is difficult to determine with precision but it must occur sometime after twenty weeks of gestation. The fact that an embryo or fetus has the potential to develop into a person does not permit us to improperly classify zygotes, preembryos, embryos, and fetuses as persons, much less "unborn babies," and treat them as such. A fertilized egg is still an egg, not an unborn chicken, and we treat it as an egg. If a single developing embryo, a human being, sometime in the first three weeks divides into two or four separate embryos, these embryos will eventually develop two or four brains and become two or four separate and distinct human beings, each with a unique personality.

Sometimes, the brain fails to develop. This is called *anencephaly*. Although anencephalics can be delivered alive, they cannot survive very long. Anencephalic newborns are human beings. They have a human genome and are living human organisms. They are not persons as we have defined the concept. Sometimes primitive spermatic or ovarian germ cells in a fetus develop into an independent mass of tissue with hair, skin, bone, nervous, and other tissues called a *ter-*

*atoma*. Teratomas can even have organs such as eyes or limbs. They are clearly human, genetically identical to the fetus in which they develop, but no one would consider them persons. This is why, when making decisions about these human beings, we accord them a lesser degree of dignity and moral respect than we accord human persons. This clear distinction makes abortion, assisted reproduction technologies, use of preembryos in stem cell production and research, disposal of teratomas, and transplantation of anencephalic organs to save other newborns morally defensible, even morally necessary, actions.

## IS GOD A PERSON?

So, what does it mean to say you believe in a personal God? Does it mean you believe that God is somehow similar to humans? Does God have self-knowledge, freedom of action, and the ability to assign value to people, objects, or events? Can God create an orderly universe, interact in a meaningful way with his creation, especially with humans, and have purposes for God's actions? This is how Collins describes his personal God. But all of these actions are brain functions. This sounds suspiciously like God being created in man's image rather than man being created in God's image. God does not have a brain. If God lacks this essential physical basis for personhood, we cannot describe God as a person.

The supernatural nature of God, should God exist, is beyond human understanding and includes God existing outside of time as a pure spirit, lacking form or material substance. God's purpose for, and means of action on, the natural material universe would be incomprehensible.

When I was a believing Catholic, I found this concept of a personal God easier to accept if part of my mental picture of God included the figure of Jesus. I accept the Bible as plausible evidence that a Jewish rabbi named Jesus lived and taught in Palestine, but I no longer accept it as valid compelling evidence that Jesus was divine. As a man, Jesus was a person with a unique personality. But believing

Jesus was also the son of God involves accepting more illogical concepts—like Jesus being both God and human simultaneously when logic demands he must have only one nature, divine or human.

According to most Christians, Jesus is not just one person but three persons in one divine person interacting with his self. Calling this trinitarian nature a mystery is only labeling it and provides no explanation at all. This is more than a mystery—it is a logical impossibility, a contradiction like a squared circle. If God exists, it seemed to me God would have to be very different from humans. As I said in my introduction, God obviously has no sex, although in keeping with the sexist tradition of most religions, most people continue to use the masculine gender when referring to God. God has no "eyes" to "see" what we do, nor "ears" to "hear" our prayers. Then why should we assume God has intelligence or a brain to think and feel as we do? I wonder what Jesus' IQ is? Does God have any parts or organs? No, God must simply be beyond human experience. God could not experience doubt or memory or make mental errors like humans.

Jesus, but not God, could experience regret, guilt, love, grief, anger, or fear in the same sense you and I do. Curiously, the Bible never describes Jesus as laughing or having a sense of humor, things commonly attributed to persons. In what sense then is God a person? On what basis do we make that assumption? On what evidence do we base this belief?

If, as Collins claims, God is "outside of time," God cannot be a person because being a person requires actions that can only occur over time. Thinking and feeling are ongoing processes that occur over time. Analyzing options, making decisions, reacting to other persons, and so on, are time-consuming operations. A divine person with an unchanging state of existence cannot think or feel. God has no mind to change, so describing God as a person is incoherent.

When I discuss God's nature and point out that God has no brain, no sex, no senses, and so on, I am not poking fun at believers. I am encouraging them to think beyond the convenient but inaccurate metaphors. I realize that a few believers will actually try to defend the concept of God as some humanlike father figure. Most believers know they are using metaphors and analogies when they try to describe a personal God. It is like "He hears us," but it's not really hearing as we experience it. "He wants us to love one another," but it's not really wanting in a human sense. Believers cannot help but give you a metaphorical description of their personal God because they cannot describe, imagine, or even comprehend the nature and characteristics of a real being that is supernatural and still accessible and intelligible to human beings. The concept of God and the supernatural is so otherworldly, so ill defined, so expansive that it cannot be condensed into a single personality in any sense of that word.

So we reduce God to terms and descriptions we can comprehend. Metaphors are figures of speech used to make a comparison using words that apply to one thing with the implication they apply to another for which they do not literally apply. Metaphorically described things may or may not actually exist. They have poetic value, not truth value, for example, the sea is not really *angry*—it is turbulent.

Jesus as a person is long dead. But the idea of God continues in our imaginations so strongly that it has assumed the appearance of knowledge or simply been incorporated into the culture as a firm belief. But humans cannot trust the things that exist in their imaginations. All imagined images are reconstructed using real-world events and objects as raw material. If I ask you to imagine a gryphon, it is unlikely you could do so. If I tell you a gryphon is an ancient word for griffin and is a beast with a lion's torso and a bird's head, you could form an imaginary picture in your head. You could imagine a beast with lion's or bird's claws with or without wings, with a head of a raven, eagle, or crowned crane. There is no way to test whether your particular image is correct since there has never been a real griffin to use as an accurate standard. You can only come up with an image if you have seen or heard of real objects that had characteristics of the imaginary one.

This is why our individual images of God are so varied. Jesus, when he walked in Palestine, probably physically resembled the Semitic people of his time with dark skin, brown eyes, and rough features and not the idealized handsome, blue-eyed, translucently white-skinned, Anglo-European figure imagined by medieval artists. Christians, when they try to imagine God, simply take parts of human nature and expand on them infinitely.

The human mind has difficulty dealing with abstraction, for example, particularly when imagining anything infinite. Trapped in a middle-sized world, we can only vaguely comprehend the extremely large or extremely small. The size of the observable universe is thought to be 78 billion light-years, and the whole universe 158 billion light-years. The size of a three-day-old embryo is about the size of the period at the end of this sentence. And a nanotube is 1/10,000th the size of one of your hairs.

This limitation of our intellects is more severe when we deal with an abstract concept such as an infinite person. How intelligent is God? God's IQ is infinite. What is God's power to control the universe? God's power is infinite. And so it goes. However, the process completely breaks down when we get to emotional mental states. How strong is God's love? God's love is infinite. How strong is God's hate? Infinite? God fears nothing but apparently has infinite anger. After considering all the evidence presented, Collins has not convinced me that conceiving of God as a person is supported by science, logic, or reason.

## IS GOD MADE IN OUR IMAGE?

I do not believe the claim that man is made in God's image simply because some ancient Jewish writers chose to write this description of mankind into their folklore texts. The biblical writers were using this metaphor to make a point about the special status of humans and their relationship with God that they wanted their sect to accept. The authors of both the Jewish Bible and the New Testament created a personality for God. Yahweh has one personality and Jesus has another.

Collins questions the concept that God created humans, using perfectly natural evolutionary mechanisms, "in the image of God" (Genesis 1:27). He points out that God does not have toenails or a bellybutton. "He" also does not have a penis or a vagina and breasts. In fact, there is not much that humans have in common with God. Collins claims this is symbolic language and refers to "a lot more about the mind than the body."[1] This seems to imply that the mind is different from and not just a part of the body.

Mind is the process produced by the activities of brain cells, just as digestion is the process produced by the cells of our digestive system. Does God have brain cells? Our brain cells produce consciousness. Does God have a function called consciousness? Humans change from one state of consciousness to another. We are not fully conscious when we sleep, when we are anesthetized, or when we are in a deep coma. Does God sleep? Is God affected by drugs or trauma? Stones are not conscious, plants are not conscious, animals are conscious to varying degrees. It seems pretty clear that God does not have a mind the way humans have a mind. So, it doesn't seem that Collins has interpreted the allegorical symbolism of Genesis correctly.

We don't want to imagine a God as just some sort of supernatural human, do we? What is the real one and only, clearly correct way of interpreting the divinely authored statements, "And God said, Let us make man in our image, after our likeness; and let them have dominion over the fish of the sea, and over the fowl of the air, and over cattle, and over all the earth, and over every creeping thing that creepeth upon the earth. So God created man in his own image, in the image of God created he him; male and female created he them" (Genesis 1:26–27)?

Scholars of Jewish history would support findings that Genesis was written at a time when the Jewish peoples were still somewhat polytheistic. The Hebrew word used by the author of Genesis for God is "Elohim." This is a plural representing the divine family of the common supreme God of the Middle East, El. This explains why God is talking to other gods. Why would God *say* anything? And if God, singular, did speak, to whom was God speaking? To other gods, it appears. One God proposes to one or more other gods that they create man in *our*, not my, likeness.

How like God are we? Does human nature share anything in common with divine nature? Knowledge? Power? Immortality? Collins and other believers need to come up with a clearer, more understandable description of just how we resemble the Supreme Being. If we are made in the image of God, is the chimpanzee 95 percent of the image of God? Perhaps it makes more sense to reverse the statement. Humankind tends to create a personal God that assumes the likeness and image of man. We attribute to God sex and vocal cords and feelings and motives that are all completely human. A purely immaterial, abstract, formless god, lacking personhood, existing outside of time and space is very different from a human being and is not the God that Collins appears to worship.

When we examine the products of human intelligence, we can logically assume certain things about the human who produced them. You can examine the universe and imagine what kind of creator might have created it, but that does not shed any light on whether such an imagined creator actually exists.

There is then an inevitable temptation to apply human qualities and nature to God with suitable exaggeration. Human products (creations) are designed for a purpose. Therefore, believers reason, using a false analogy, that the universe as a product of creation must have an intelligent, purposeful designer. The conclusion need not necessarily follow. Purpose only entered the universe when the human brain developed the concept, and purpose only applies to things we do. While it is usually easy to determine the purpose of products created by humans, the universe itself has no discernible purpose except to exist as a universe.

Believers have used the argument that the order and design seen in the universe demands a divine designer, namely, God, for centuries.

One of the most influential proponents of the argument was the Reverend William Paley (1743–1805) in his book *Natural Theology or Evidence and Attributes of the Divinity Collected from the Appearances of Nature.*[2] He begins his discussion by describing a walk across the heath where he first kicks a stone and then finds a pocket watch. The famous pocket watch found by Reverend Paley while walking on the heath in 1802 is the basis of an argument for an intelligent "person" behind the design of the universe. Paley's use of a flawed analogy, that a pocket watch is a mechanical device designed to meet a human need and therefore the universe is just a mechanical device to meet God's need, invalidates his argument.

The valid conclusions that Paley could have drawn about the watch were (1) that it was an artificial device, that is, not naturally occurring without human intervention; (2) that its production was a complex operation requiring knowledge of natural processes and natural laws of motion; and (3) that it was designed by a person to measure time.

Paley could not have found such a watch in 1400 CE when strolling the heath because clocks at that time were much simpler devices that used water, weights, and gears to measure time in hours. The history of the development of the pocket watch includes humankind's interest in measuring time for various practical purposes going back to Egyptian shadow measurements. The pocket watch *evolved.* The first watches were not as complex as modern watches. It was not created in an instant out of nothing by a brilliant designer with the intention to know what time it was.

Paley used his mind to reason, by analogy, that if all things that are machines are designed by intelligent human persons, the universe was also a machine and therefore had to have an intelligent designer and a purpose. But Paley did not establish that the universe is a machine. He, in fact, also kicked a stone during his stroll and thought it was unremarkably natural and part of creation without purpose or design. It gave him no evidence of a personal creator or intelligent designer because the rock was not a machine or human invention.

The universe contains many "designs," ordered regularities, that produce inefficient, cruel, and ugly results that reflect badly on the intelligence or intentions of its supposed designer.

What justified Paley's assumption that what applies to artificially produced human machines also applies to naturally occurring objects and organs? Starting with simple building materials and following a few simple laws, nature can over the course of time produce very complex objects and organisms. What else could Paley know about the designer of the pocket watch? Could he tell whether the designer's purposes and intent were good or evil? Was the designer an angry, revengeful, merciless man, or a passive, content, forgiving man? The point is that only limited, objective personal knowledge can be gained about the designer's nature simply by looking at products he designed. Nor can it be assumed that the designed product was very good at achieving the purposes the designer intended for it. How accurate was the pocket watch Paley found? Was it accurate to the hour, minute, second, nanosecond? Did it function at all? The regularities and patterns we find in nature are inappropriately equated with man-made machines and used to establish the existence of a divine designer. For a detailed refutation of the argument from the seeming design of the universe, see Richard Dawkins's classic *The Blind Watchmaker*.[3]

Even if we wanted to agree with Paley, it is another longer leap of logic and faith to identify his supposedly intelligent designer with the person of Jesus Christ. According to Collins, the launching pad for that leap is the Bible's New Testament and the C. S. Lewis book *Mere Christianity*.

## WHY BELIEVE GOD IS A PERSON?

If God does indeed exist, Christians, like Collins, have only two alternatives. Either god is an impersonal spiritual force that created and sustains the natural universe and has no special relationship or concern for humans past, present, or future, or God is a person who interacts with humans, desires their worship and fellowship, and will reward or punish their behavior based on a set of moral laws. Clearly, Christianity makes no sense without a personal God.

The dictionary defines "fellow" as a comrade or an equal, one of a group having common interests and characteristics. How can we have fellowship with God if God is not one of the fellows? Do you desire fellowship with monkeys or baboons? No, we do not have enough in common. What does Collins mean when he asserts his belief that God desires fellowship with humans and no other creatures?

Where is the evidence that would support the concept that God was a person or even render it plausible? The more I look for evidence, the more the evidence leads me away from a personal God to an impersonal, mysterious, abstract spiritual power beyond human comprehension. I did not read *Mere Christianity* until after much study and contemplation had led me to the conclusion that God was not a rational possibility. The foreword of the edition I read accurately stated, "The 'mere' Christianity of C. S. Lewis is not a philosophy or even a theology that may be considered, argued, and put away on a shelf." The book certainly is not a closely reasoned, unambiguous recitation of evidence supporting Christianity as the one true belief. It is full of irrelevant analogies, uninformative metaphors, and rhetorical flourishes characteristic of C. S. Lewis, the skilled author of fairy tales like *The Lion, the Witch, and the Wardrobe*. C. S. Lewis was in love with his imagination, creating fantastic alterative worlds using emotionally charged words, sentences, and paragraphs. There is a whole body of literature criticizing C. S. Lewis and analyzing his work from philosophic, scientific, and psychological points of view.[4] I found Lewis's writing extremely wordy, difficult to read, and not very convincing.

What Collins and Lewis offer in support of their belief in Christianity can be summarized by Blaise Pascal's phrase, "The heart has its reasons that reason cannot know." Faith, acceptance based on intuition, desire, longing, or feeling of incompleteness—not rational, intelligent review of the quantity, quality, and relevant evidence and experience—is the ultimate basis of their beliefs.

In fact, Collins has cited no scientific evidence that would support the concept of God as a person. Nothing supernatural is required to define and characterize human personalities or their development. Christians, like Collins, have to resort to the poor and unreliable "evidence" of the Bible stories to defend their belief that God is a person. Given the history of the Bible and the biased perspective of its unscientific and relatively ignorant authors as detailed in chapter 6, this is no evidence or reason to accept the existence of a personal God.

## DOES GOD HAVE MULTIPLE PERSONALITY DISORDER?

My belief in Christianity was further impeded by the fact that the personal God of the Hebrew Bible seemed to be a completely different person from the God of the New Testament. The God of the Hebrew Bible, Yahweh, is very much concerned with making a covenant with the Jewish people and dealing with the enforcing, breaking, and remaking of this covenant. He is a merciless God, who inflicts pain, death, and misery on innocent men, women, and children in the cause of establishing his holy land of Israel. Karen Armstrong, a former Catholic nun who writes about comparative religion, correctly describes the God of the Hebrew Bible as follows: "This is a brutal, partial, and murderous god. A god of war, who would be known as Yahweh, Saboth, the God of Armies. He is passionately partisan, has little compassion for anyone but his own favorites, and is simply a tribal deity."[5] She disputes the theory that all the major monotheistic religions—Christianity, Judaism, and Islam—worship the same God of Abraham, and she describes how our name for and view of the nature of God changed over time to suit the need of changing culture and civilizations.

Yahweh was apparently unwilling to grant the gift of free will, as he frequently intervened to force humans to behave in specific ways. Yahweh is often "hardening the heart" of various people against the Jews (Exodus 4:21; 7:3, 7:13; 10:1; 10:20; 10:27; 14:4; 14:8; 14:17; Deuteronomy 2:30). In Genesis 20:6, God withholds Abimelech from sinning. In Judges 9:23, God sends an evil spirit between Abimelech and the men of Shechein. Moses says in Numbers 16:28, "Hereby ye shall know that the Lord that sent me to do all these works, for I have not done them of mine own mind." And the Lord put a word in Ballam's mouth (Numbers 23:5). "I will send a faintness into their hearts" (Leviticus 26:33).

Karen Armstrong traces the evolution of Yahweh from a God, utterly incomprehensible, quick to anger, and terrible in his wrath, preoccupied with the minuscule detail of the lives of his chosen people and unconcerned with the Gentile world, to a God that encouraged "compassion and respect for fellow human beings." Yahweh was a lawmaker, and his 613 laws remain part of the infallible Bible.

This seemed to me very different from Jesus, who forgives sin, heals the sick, and offers the Golden Rule. How can these different visions of a personal God be combined in one nature? *The Language of God* and the writings of C. S. Lewis do not address the whole issue of the different personalities of personal Gods. As I read the religious literature of Islam, Judaism, Taoism, Hinduism, and so on, the humans in these cultures seem to have deep commitment to an incompatible variety of different divine personages.

Furthermore, believing as I did that God has a divine providential plan for the universe and the humans in it, I saw everything that happens as a manifestation of God's will. Having a will, purposes, and intentions and acting accordingly is one of the criterions of being a person. But if God's personal will is always done, where does that leave the concept of free will? Were all my prayers and good works only the result of God controlling my mind and my experience, leaving me with the illusion that I had a possibility to do otherwise?

Again, God does not seem to meet the criteria of personhood. A God that loves good and hates evil so much that he would never allow evil to occur would make sense if God took charge and God's will determined everything. How do I reconcile this to the belief that God gives me the gift of free will and lets me earn salvation or damnation? It was impossible to be intellectually rigorous and honest and still maintain my belief that Jesus was a personal God that made sense.

Once the concept of a personal God is lost, all that remains is the concept of the god of the philosophers as first cause, creative spirit, and so forth. At this point, you lose the concept of religion with its stories, myths, rituals, doctrines, and dogma. There is no way to communicate with an impersonal god, and even if such a god exists, it is irrelevant to humans because it does not care what they do during their brief lives.

## NOTES

1. F. Collins, *The Language of God* (New York: Simon & Schuster, 2006), p. 205.

2. W. Paley, *Natural Theology or Evidence and Attributes of the Divinity Collected from the Appearances of Nature* (1802; Oxford: Oxford University Press, 2006).

3. R. Dawkins, *The Blind Watchmaker* (New York: Norton & Company, 1987).

4. Critical analyses of C. S. Lewis include D. Baker, *Losing Faith in Faith: From Preacher to Atheist* (Madison, WI: Freedom from Religion Foundation, 1992); J. Beversluis, *C. S. Lewis and the Search for Rational Religion*, rev. ed. (Amherst, NY: Prometheus Books, 2007); S. T. Joshi, *God's Defenders: What They Believe and Why They Are Wrong* (Amherst, NY: Prometheus Books, 2003); D. Holbrook, *The Skeleton in the Wardrobe: C. S. Lewis's Fantasies, a Phenomenological Study* (Lewisburg, PA: Bucknell University Press, 1991).

5. K. Armstrong, *History of God* (New York: Knopf, 1994), p. 19.

# CONCLUSION

*Happiness is nonetheless true happiness because it must not come to an end, nor do thought and love lose their value because they are not everlasting.*

—Bertrand Russell

## SUMMING UP

My main purpose in writing this book has been to question the claim made by Collins in the introduction to *The Language of God*:

> So here is the central question of this book: In this modern era of cosmology, evolution, and the human genome, is there still the possibility of a richly satisfying harmony between the scientific and spiritual worldviews? I answer with a resounding yes! In my view there is no conflict in being a rigorous scientist and being a person who believes in God who takes a personal interest in each one of us.[1]

Collins's book is certainly not embraced by fundamentalist, Bible-believing, young-earth creationists or by their hypocritical, pseudo-scientific, Bible-believing imitators who are trying to sell intelligent design. However, the book is still used by moderate religionists to defend their particular belief in a personal God as rational and consistent with modern science. They use the great and deserved respect paid to Francis Collins as a scientist to bolster their position on religion. Contrary to Collins's claim that many scientists share his view, such scientists are rarities. Apparently one of Collins's reasons for writing his book was to protect religious believers from the religion-haters' charge that their beliefs are stupid, irrational, and antiscience. Collins attempted to provide them with a theistic defense of what modern science accepts as the origins of the cosmos and evolutionary development of humankind. A large portion of his book is therefore devoted to a reinterpretation of Genesis and a review of the scientific evidence consistent with evolution. While he does an acceptable job of reviewing the scientific evidence for evolution, emphasizing his knowledge of genetics, there are better secular books devoted specifically to the subject.[2] I have no quarrel with the evolution part of theistic evolution.

However, when it comes to the theistic belief in a personal creator God, Jesus Christ, who hears and responds to prayers and miraculously intervenes in the universe to achieve fellowship with individual humans, I find Collins's thinking magical and his arguments unpersuasive. His arguments are more reflective of the needs of Collins's unique human personality than of his scientific intelligence.

My intention is to provide some sound evidence and observations to support my critique of the belief in Jesus as personal God and to put the issue of science versus the Christian religion in reasonable perspective. My intention has also been to offer an alternative to the often mean-spirited criticism of religion that some of the more popular "God and religion haters" use to defend the scientific atheist view.

While Collins accepts that belief in a deist god is rational, he rejects this kind of god because, by definition, this is not a god with whom humans could interact. I did not emphasize or exhaust the arguments against the existence of such an impersonal deity. I touched on but did not fully discuss the scientific analysis of free will or consciousness or the natural reasons why religions evolved and persist as social institutions. Finally, neither Collins nor I have discussed other versions of personal gods predominant in non-Western religions, such as Islam, Hinduism, Taoism, and so on. Adherents of these religions might propose different challenges to Collins's selection of Jesus Christ as the one true God.

For Collins, "The Moral Law still stands out as the strongest signpost to God. More than that, it points to a God who cares about human beings and a God who is infinitely good and holy."[3] By "Moral Law" Collins means that it is natural for each human being to establish for himself a standard for ideal behavior with respect to other human beings. I agree that most human beings are naturally constituted to do this and that we have genetically produced basic instincts that lead us to build our own unique, internal mental standards of behavior. From infancy on, these images of behavioral ideals are reinforced, modified, and strengthened by our caretakers and kin. They become ever more sophisticated and elaborate, covering ever more specific sets of circumstances. Our environment also provides experiences that mold our developing moral standards. This process appears entirely natural and necessary, not only for humans but also to a lesser extent for all social animals.

The voice of conscience that Collins hears regarding moral issues is, in reality, the same voice that he hears when reading a scientific journal or figuring out how to find a specific street address. It is the result of active synaptic pathways in his brain and reflects conscious awareness of his behavioral guides drawing on past experiences that have been incorporated into his personality. Were he under pressure from extreme fear, anger, depression, fatigue, drugs, starvation, toxins, or physical injury, for example, both his moral standard and behavior could easily change.

The overwhelming scientific evidence that material things have direct influence in creating the Moral Law makes it clear that there is nothing supernatural about it. The moral sense has a natural, evolutionary origin. Psychiatrists and neurocognitive scientists have identified numerous people who have little or no sense of Collins's Moral Law. Chances are, if you watch any of the many crime shows on television, you have probably seen one of these sociopaths interviewed. Courts excuse horrendous behavior based on the accused's inability to distinguish right from wrong. What evidence does Collins cite to support even a suggestion that this mental capacity comes from God? None. In fact, he arrives at his conclusion that the Moral Law is the instillation of a glimpse of divinity only by denying the evidence for the causal role of genes and custom. This leaves the unproved, default explanation that "Jesus did it." You don't prove your explanation is true by proving some other explanation is false or incomplete.

The Golden Rule did not originate with Jesus Christ nor is it unique to Christianity. The Jews did not reject Jesus and the Romans did not kill him because he taught the Golden Rule. The fact, if it is a fact, that he taught it does not constitute evidence that Jesus was divine nor does it qualify him for personal worship. The Iranian prophet and religious poet Zoroaster, after a vision from the god Ahura Mazda in about 650 BCE, founded the first recorded religion that had an essentially monotheistic creator god. This god established moral rules for humans to use in battle with evil forces and demigods and would send a savior in an apocalyptic end time to judge men. Zoroaster's god would assign humans to heaven or hell based on their observance of their version of their "Golden Rules." Versions of the Golden Rule existed in Chinese Confucianism and Hindu scriptures before these religions had any contact with Christianity. Intellectuals of ancient Greece and Rome, the atheists of the Enlightenment, and secular humanists today, all quite independent of God and religion, have espoused the Golden Rule and moral behavior.

For thousands of years humans saw their gods as basically unconcerned with the moral behavior of humans. Gods were concerned with controlling events in nature, crops, childbirth, floods, earthquakes, and the like or personal popularity, politics, wars, competitions, or prosperity. There was no sin or afterlife to reward or punish unethical behavior. Philosophers and kings, using practical reason as their guide, decided ethics and morality. Still, there were good and moral Greeks and Romans and Egyptians. So it is clear that you can be good without appealing to God as the source of morality. Who has the moral high ground—an atheist who is good knowing that no one is aware of his good deeds, or a believer who believes God is always watching, ready to punish or reward?

The other argument that Collins finds compelling is the anthropic principle, which states that so many physical constants have the values that are conducive to the evolution of intelligent life and that the God of the gaps is the only probabilistic cause. This issue has been addressed in chapter 5. But even if a divine intelligence were responsible for the constants, there is no way to establish that this deity was Collins's personal God, Jesus. The only route to Jesus as the personal God of Collins's dream is the Bible. The weakness of this source as a means to demonstrate Jesus' divinity has been clearly demonstrated in chapter 6. My verdict is the one permitted to juries under Scottish law: "not proven." Moreover, it is not even plausible.

## EXHORTATION TO BELIEVERS

It is my hope this book will move you to honestly examine the reasons you hold your beliefs about God. How much do your beliefs depend on your trust in others rather than on personal examination of the evidence? Do you continue your participation in religion to avoid hurting your family or conflict with your friends? Does your love for your parents require that you accept without question the marriage partner, vocation, or political party your parents selected for you? Why should you accept their religious beliefs?

To what degree do your beliefs depend on a Bible interpretation by someone who claims to have accurately determined the word of God? Do you "know" that Jesus is God because in a time of desperation he answered your prayer? Considering the number of desperate people who pray just as fervently and fail to get any help, could it be that you were just lucky? Are you concerned about saving your immortal soul from hell? Even our current limited scientific understanding of conscious brain function and how the human mind operates provides a natural explanation of what the Christians call the soul. We have consciousness but no immortal soul. I was not conscious for billions of years before I was born, and when I die, I shall again be unconscious for eternity. As put by David Hume, famous Scottish philosopher, "The Soul, therefore, if immortal, existed before our birth, and if the former existence in no way concerned us, neither will the later." My vision of death is lying in a deep, dreamless sleep or being deeply anesthetized and never waking up. I do not fear hell.

Believers demand respect for their beliefs and rituals no matter how bizarre or illogical they may be. They accept overt and covert attacks on other religions, pagans, infidels, Jews, and so on, but want their own priests and ministers to be free from outside questioning and criticism. Have you ever heard of believers insisting on equal time for "teaching the controversy" between atheism and a personal God? I know that people can find comfort in false beliefs. Many times I have heard the well-intended lies, "You don't have cancer," "Things are bound to get better," "Her death was all for the best." False beliefs do not change reality, whatever their benefit.

As a Christian, examine your attitudes and behaviors toward non-Christians. Are Islamic and Jewish beliefs, about who Jesus was and what he said and did based on their interpretation of their own divine revelations, more or less believable and acceptable than yours? How do you test the accuracy and source of a possible divine revelation? All religions are select groups. The Jews have their Gentiles, the Muslims their infidels, and Christians their pagans. Every believer accepts his particular religion or worldview as the only true religion. The Jews deny that Allah and Jesus are divine. The Christians deny Allah and Yahweh are divine. They do not worship and obey the same god. How can they all be true? Do you fear, hate, refuse to associate with, or ignore people who are atheists, scientists, homosexuals, Jews, sinners of every kind? Atheists routinely receive a significant amount of angry hate mail using foul language that reflects the depths of the emotions of closed-minded Christians. I do not hate God or love the devil any more than I hate or love the tooth fairy or leprechauns or gremlins. They are simply irrelevant. I urge you to develop tolerance and open-mindedness and, at the very least, use reasonable dialogue to explore your differences.

Refrain from advocating the use of the police powers of the state or public funds to impede scientific and medical progress or to impose your beliefs on others as long as the state tolerates and permits you the free personal expression of your religion. The public sphere should be free of unnecessary laws based on one religious "moral" point of view or superfluous religious rituals or references falsely implying a universal belief in God. Respect the freedom of each individual to use his or her own intelligence to make sense of life experiences and to find answers to the questions I have raised in this book.

Science is an international institution with time-tested methods that include careful, detailed observation and measurement, critical analysis of experience and experiment, and an obligation to constantly check and refine theories. Scientific knowledge has resulted in real, explanatory, productive, and practical impact. It is pointless and futile to ignore or reject the information and technologies that science provides. Study science and religion. Deepen your knowledge about their history and interactions.

In any conflict between science and religion, history shows that science ultimately triumphs. A Gallup Poll conducted in December 2008 found that only 53 percent of adults in the United States believe that religion can answer most problems. This represents a steady loss of confidence in religion since 1958, when 82 percent shared this sentiment.[4] Science, as we know it, has been active for less than five hundred years, while Christianity has been around two thousand years. God is not dead but is slowly dying.

We need a better source to serve as the ground for our moral behavior. If you talk to atheists or agnostics, you will find they are no more or less moral than Christians. Calling someone a godless atheist (a redundant epithet) is not the same as to declare the person evil or untrustworthy. The majority of atheists are not obnoxious, angry, ignorant, or unhappy people. Shouldn't people be characterized by what they do and don't do rather than by what they do or don't believe about the supernatural? Compliance with a divinely instilled Moral Law as interpreted by some religion's clergy and constantly monitored by an all-seeing God is not the only source of decent, ethical behavior. Some are motivated to be good by the bribe or reward of heaven, others by the fear of hell. In contrast, humanism is a worldview that supports a strong set of ethical principles including respect for human dignity and rights, honesty, justice, compassion, and concern for the sick, poor, old, disabled, and disadvantaged. Humans perform simple acts of unobserved, unrewarded kindness all the time, whether they are atheists, agnostics, or members of a non-Christian religion. Humanists find a personal purpose and infuse their lives with meaning rather than accepting or searching for meaning and purpose in an imposed divine plan.

In multicultural New York City, was everyone who rushed to the aid of the 9/11 victims a Christian or even religious? I have noticed that in grief, caring human beings comfort us, and they need not have a clerical collar or wear a yarmulke. When a fellow human is in danger, people, not angels, help regardless of their religious beliefs. We need to respect each other's dignity and do what ultimately results in what is best for all concerned given the circumstances. We need to avoid or minimize what would increase physical or mental pain.

Last, I would like to address "Pascal's wager." Blaise Pascal (1623–1662) was a mathematician who worked on probability, among other things. He proposed that living a religious life was a gamble. If God exists, when you die, you reap a great benefit—heaven. But if God does not exist, Pascal claims you have lost very little. However, if this life is your one and only chance for any sort of happiness or pleasure, you will have sacrificed everything for nothing at all. What God should you bet on—Jesus, Allah, Yahweh, or Vishnu? Hope is the most necessary and yet can be the cruelest of virtues. Hope that you will receive the happiness in the next world, that all injustices will be corrected, and that you will be united again with loved ones is vain. Hope in a brighter future here on earth is essential.

## EXHORTATION TO SCIENTISTS

Contrary to Collins's assertion, atheism or agnosticism appears to be the most rational worldview. This universe looks exactly the way one would expect if it consisted of matter and energy changing in accordance with rules that are only gradually becoming apparent to humanity. Deism (an impersonal god) and pantheism (god existing as a spiritual consciousness infused throughout the universe) are wrong but more acceptable and consistent with nature and its laws. They do not lead to the frequent excesses, abuses, intolerance, and misery caused by evangelistic belief in a personal God.

Scientists must stop being intimidated by the popular acceptance of religion and courageously and compassionately confront the use of blind, irrational, Bible-based faith to influence the lives of humans throughout the world. Christians' "moral" reservations should not be allowed to immorally restrict research involving embryos, contraception, abortion, evolution, environment, and the treatment and prevention of sexually transmitted disease. Scientists should accept the moral responsibility to see that scientific knowledge is used to promote pleasure, comfort, and security and not contribute to anxiety, worry, pain, and war.

Religion's concerns should be acknowledged and considered but should not be determinative in our secular government's policy decisions if such concerns are against the common good. The public sphere should be free of unnecessary religious references and rituals out of respect for the diversity of views on the subject. Scientists need to promote critical thinking and aggressively push for better science education. Scientists should be active in promoting a humanist agenda that involves working with and for the citizens of their communities. Scientific humanists must show that religion is not necessary to live a full, happy, and productive life. Scientists should strive to retain their objectivity and resist popular opinion and the temptation of money to reach accommodation with any specific religious agenda. The fact that most prominent believing scientists have received the lucrative Templeton Prize raises the suspicion that they were consciously or unconsciously influenced in their thinking.

Scientists who defend their nonreligious worldview need to avoid elitist tactics such as denigrating the intelligence of their religious opponents. They should make every effort to communicate with the nonscientific community by using understandable language and avoiding technical jargon. When initiating a dialogue, do not assume that your audience knows the basic concepts of science but willingly and patiently educate them, building on information they already know.

Remember, you are not trying to persuade a believer to change his brand of toothpaste; you are asking him to change his whole view of life. If he adopts your perspective, his life in this world will no longer be a temporary period of trial, a journey to what he believes is the real supernatural world intended for us, where all of our worldly problems will be resolved. No, as a naturalist you are telling him that we live in this world and are of this world. The journey begins and ends in this world, as imperfect as it is, and some problems are never completely resolved. You are telling him that he will have to figure out how to make this the most pleasant journey possible because this is all he has. But be forewarned: although you can win some converts, many will not have the ability to make such a drastic change in their worldview.

## COMMITMENT TO TRUTH

In the movie *A Few Good Men*, Jack Nicholson, playing the role of a Marine commander on the witness stand, is trying to conceal or justify the murder of one of his men. In response to Tom Cruise's relentless questioning, he finally breaks down and blurts out, "You want the truth? You can't handle the truth!" I wonder how many of you can handle the truth if the truth is uncomfortable or scary. It sometimes takes a lot of courage to face the truth. Are you willing to seek the truth whatever the consequences? Are you willing to accept and act on an inconvenient truth? Curiously, both Collins and I are asking this same question. From Collins's point of view, atheists have a hard time accepting the truth that we have to give up our sinful pleasures, our self-centered, selfish, materialistic, immoral lifestyles, and pay attention to the "new requirements on your life plans" that God imposes. In his opinion, atheists need to learn to endure the inevitable suffering and injustices of this life. As compensation, Collins offers the imaginary love of an imaginary friend in the sky and a better life after your death where you are united with God and your dearly departed in heaven.

The hard truth that I have accepted and that Collins has not, is that there is no supernatural reality. My dearly departed are forever gone. While suffering and injustice are inevitable, our life plans should actively try to minimize them and maximize pleasure for ourselves and our fellow man. The compensation for accepting the truth of a scientific worldview is the joy of the physical pleasures of the body and the intellectual pleasures of an active mind. Essayist Debra Dickerson articulated it beautifully: "The world can take everything away from you but your mind. I never aspired to prominence or money (though I am cool with both), just intellectual fulfillment and the capacity to master whatever I choose, wherever it led."[5] For science, the ultimate value is truth; for religion, the ultimate value is unquestioning faith.

I would like to suggest a test to help you decide whether your current worldview is working for you. When I read the works of German social critic Friedrich Nietzsche (1844–1900), I came across his idea of the Eternal Recurrence. This concept, which he developed from its origins in Eastern cosmology, is the view that reality consists not of a straight-line progression, but of ever-repeating cycles. He offers the test in the form of a question. If you could choose to live your life over from beginning to end, without any changes, and then repeat the experience again and again, would you accept the offer? My answer is a resounding yes. I hope your answer is also positive. Of course, we would all like a second chance at life, provided we could improve it based on what we learned the first time around, but that is not Nietzsche's test.

I hope that at this point I have convinced you that science is the only source of true knowledge about the real natural universe and that there is no reason to believe in a supernatural reality. Science cannot be harmonized or reconciled with religious faith. Collins, or anyone else, can use the tools of science to contribute to our knowledge of the natural world, but they cannot use them to support the theory that God or the supernatural exists. My debate with Collins has been conducted to stimulate the parts of our brains that deal reasonably with evidence collected by experience in the real, external world. But there are other parts of our brain that can experience another world that can seem every bit as real. This is a world that we create internally in our minds, which has no external counterpart. This personal, subjective world is not open to challenge by reasonable argument or by reference to external reality.

Religious beliefs based on dispassionate logical analysis of evidence and arguments can be challenged and changed, but religious belief based on powerful mental experience cannot be affected by scientific evidence, logic, or reason. A clear demonstration of the difference is recounted in the book *The Spirit Molecule* by Rich Strassman.[6] Strassman, a clinical psychiatrist, studied the effect of a mind-altering chemical found in the brain called dimethyltryptamine (DMT). He administered a small amount of DMT into human volunteers intravenously. Within a minute or two of injection the subjects had a strong psychedelic experience that lasted for about five minutes. Subjects could talk in about fifteen minutes and were normal in thirty minutes. A variety of experiences were reported including visual and auditory hallucinations, out-of-body experiences, and various emotional states. Interestingly, one of the most common reports was of being in contact with other beings, such as space aliens, and interacting with them. Interactions included the aliens using them in experiments and reprogramming their brains. The subjects accepted their experiences as real-life events even though these experiences were clearly only brief responses to the injections.

Strassman reported:

> The research subjects tenaciously resisted biological explanations because such explanations reduced the enormity, consistency, and undeniability of their encounters. How could anyone believe there were chunks of brain tissue that, when activated, flashed encounters with beings, experimentation, and reprogramming? Neither did suggesting that it was a waking dream satisfy the volunteers' need for a model that made sense and fit their experience. Many even prefaced their reports by saying, "This was not a dream," or, "I couldn't have made it up if I wanted to."

While this is a "true" experience, this is not an experience of the truth of the real world.

Ultimately, I hope that I have convinced you that belief in God and the supernatural is based entirely on faith without evidence. The fact that Collins accepts both scientific knowledge and the unsupported belief in God is an example of cognitive dissonance; it is not, as Collins and his admirers' claim, a harmonization of contradictory worldviews.

## NOTES

1. F. Collins, *The Language of God* (New York: Simon & Schuster, 2006), p. 5.

2. R. Numbers, *The Creationists: From Scientific Creationism to Intelligent Design* (Cambridge, MA: Harvard University Press, 2006); S. B. Carroll, *Endless Forms Most Beautiful* (New York: W. W. Norton, 2005); S. B. Carroll, *The Making of the Fittest* (New York: W. W. Norton, 2005); D. Futuyma, *Evolutionary Biology* (Sunderland, MA: Sinaver Association, 1997).

3. Collins, *Language of God*, p. 218.

4. http://www.gallup.com/poll/113533/Americans-Believe-Religion-Losing–Clout.aspx (Accessed January 24, 2009).

5. D. Dickerson, Riff blog, http://www.motherjones.com/riff/2007/12/me-mommy-dearest.

6. R. Strassman, *The Spirit Molecule* (Rochester, VT: Park Street Press, 2000).

# INDEX